S0-AVR-919

The Earth After Us

The Earth After Us

What Legacy Will Humans Leave in the Rocks?

Jan Zalasiewicz

WITH CONTRIBUTIONS FROM

Kim Freedman

OXFORD
UNIVERSITY PRESS

Great Clarendon Street, Oxford OX2 6DP

Oxford University Press is a department of the University of Oxford.
It furthers the University's objective of excellence in research, scholarship,
and education by publishing worldwide in

Oxford New York

Auckland Cape Town Dar es Salaam Hong Kong Karachi
Kuala Lumpur Madrid Melbourne Mexico City Nairobi
New Delhi Shanghai Taipei Toronto

With offices in

Argentina Austria Brazil Chile Czech Republic France Greece
Guatemala Hungary Italy Japan Poland Portugal Singapore
South Korea Switzerland Thailand Turkey Ukraine Vietnam

Published in the United States
by Oxford University Press Inc., New York

First published 2008

British Library Cataloguing in Publication Data
Data available

Library of Congress Cataloging in Publication Data
Data available

Printed in Great Britain
on acid-free paper by
CPI Antony Rowe, Chippenham, Wiltshire

ISBN 978–0–19–921497–6

1 3 5 7 9 10 8 6 4 2

To my parents, and to the
late John Norton of Ludlow Museum.
They provided the start.

Contents

Acknowledgements

This book has been an unconscionably long—almost geological—time in the writing. I'd like to thank, first, Gabrielle Walker, then at *New Scientist*, who encouraged an early essay on this theme (also written with contributions from Kim Freedman, who has a rare skill at bringing palaeontology to life). Gabrielle encouraged further forays into this kind of writing, as did Jimmy Leach at the *Education Guardian* and Phil Donoghue at the Paleontological Association. Latha Menon's editing of this book at OUP was done with great skill and tact (gently removing those baroque sections that I would otherwise have deeply regretted subsequently); the overall shape of this book owes much to her. Her colleagues at OUP (too numerous to mention individually: I little realised how complex a business is the publishing of a book) were likewise all a pleasure to work with.

The whole or parts of this book have been read by my colleagues including Roy Clements, Peter Friend, John Hudson, Adrian Rushton, Alan Smith, Alex Page, Kip Jeffrey and Ryszard Kryza, while Andy Gale also gave advice on an early draft of one particularly intricate section. Their corrections of my unforced errors, and suggestions for additions and amendments, were invaluable—though they hold no responsibility for the content, and especially for the more speculative parts of it. The blame there is mine alone. The idea was to explain the workings of the science of stratigraphy through the future of humankind and of the fruits of its industry. Whether or not this has worked will be for you, the reader, to decide.

More generally, my writing was shaped through the tradition of repeated editing at the British Geological Survey. There was Tony Bazley's precision and patience in my early days, for instance; one gets it right, eventually. Adrian Rushton, peerless in the infinitely complex world of stratigraphical palaeontology and also in finding the *mot juste*, was endlessly encouraging, as was Tony Reedman in wider aspects of geology. As regards the science itself, my colleagues down the years—perhaps in particular those with whom I have investigated rocks old and young and those of the Stratigraphy Commission of the Geological Society of London—have provided me with an indispensable and urbane education in this most misunderstood of subjects. So too have the colleagues—Mike Branney, Sarah Gabbott, Mark Williams *et al.*—with whom I work at the University of Leicester's Department of Geology. Elsewhere, Barrie Rickards has been a mainstay of graptolite science for me and many others, Ryszard Kryza has been a marvellous guide to these rocks that have endured far too much history, Karel de Pauw has continually kept me informed about science beyond geology, while the late Harry Leeming's influence as regards scholarship *sensu lato* was profound.

My wife Kasia and son Mateusz bore the time-devouring monster that is book-writing with great fortitude, and more (Mat's input into the cover picture, for instance). Crucially, they gave me the gift of a whole, uninterrupted summer's month in which I could finally wrestle this thing to the ground. I hope that, to them, it was worth it.

GEOLOGICAL TIMELINE

billions of years from present

0
-1
-2
-3
-4

PHANEROZOIC
'PRECAMBRIAN'

?
Cenozoic
Mesozoic
Palaeozoic
PROTEROZOIC
ARCHAEAN
HADEAN

Meeting point with the aliens of this book
Humans evolve
Dinosaurs become extinct
Life invades land
'Cambrian explosion' – origin of all major animal groups
'Snowball Earth' glaciations
abundant microbial life in sea
first supercontinent?
beginning of oxidation of land surfaces
major oxygenation of oceans – large-scale Banded Iron Formations
abundant microbial life in sea
oldest microfossils
photosynthesis begins
oldest rocks of Earth's surface
? life begins
oldest terrestrial crystals
Moon-forming impact
4567 – formation of Earth

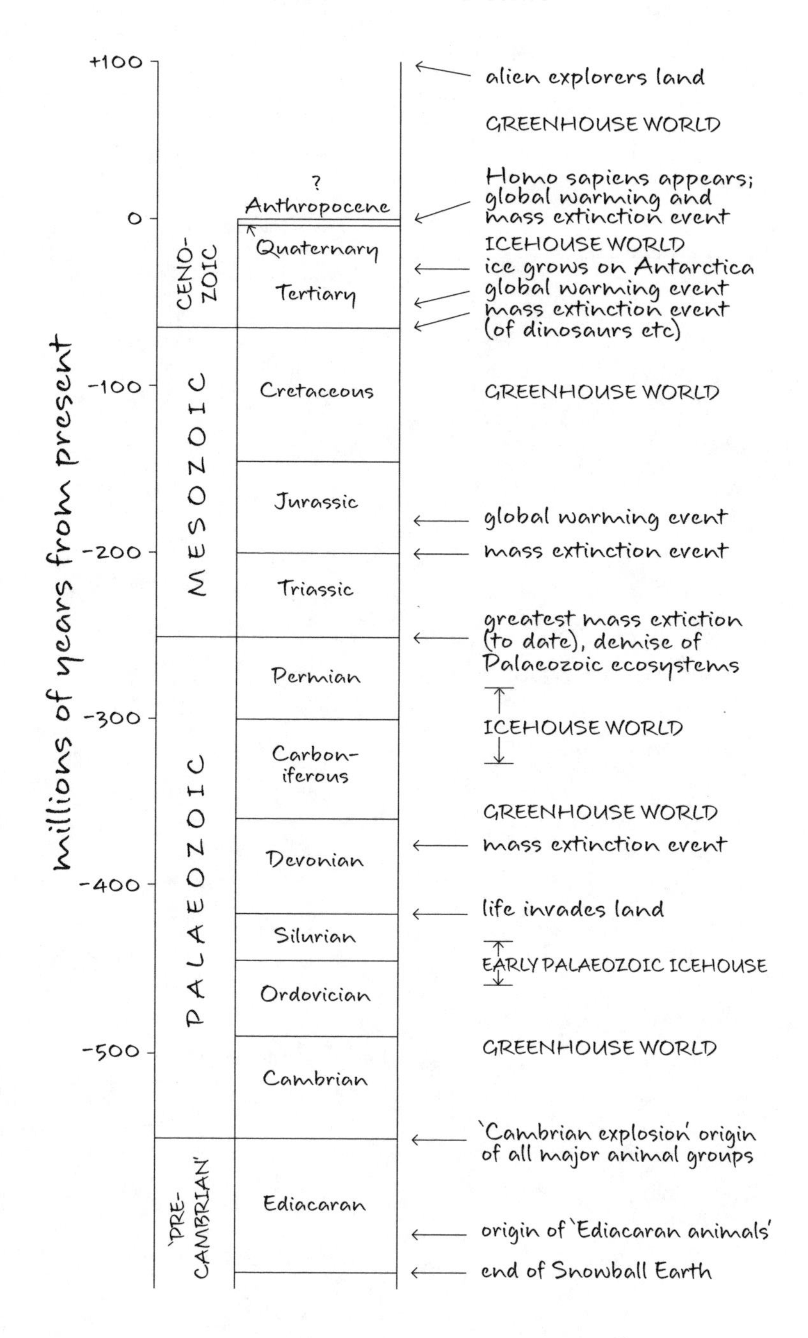

+100
0
-100
-200
-300
-400
-500
millions of years from present
CENO-ZOIC
MESOZOIC
PALAEOZOIC
'PRE-CAMBRIAN'
?
Anthropocene
Quaternary
Tertiary
Cretaceous
Jurassic
Triassic
Permian
Carbon-iferous
Devonian
Silurian
Ordovician
Cambrian
Ediacaran
alien explorers land
GREENHOUSE WORLD
Homo sapiens appears; global warming and mass extinction event
ICEHOUSE WORLD
ice grows on Antarctica
global warming event
mass extinction event (of dinosaurs etc)
GREENHOUSE WORLD
global warming event
mass extinction event
greatest mass extiction (to date), demise of Palaeozoic ecosystems
ICEHOUSE WORLD
GREENHOUSE WORLD
mass extinction event
life invades land
EARLY PALAEOZOIC ICEHOUSE
GREENHOUSE WORLD
'Cambrian explosion' origin of all major animal groups
origin of 'Ediacaran animals'
end of Snowball Earth

Prologue

The deepest of the canyons that cut through the mountains of the Great Northern Continent finally provided the solution to the riddle. The expedition team, picking their way through the boulders at the foot of that mighty ravine, knew, even as they first glimpsed the rock layer, that this might be what they were looking for. The stratum, tilted at a crazy angle by the earth movements that had thrown up the entire mountain range, was . . . different. Metres thick, with irregular protrusions, an irregular patchwork of grey and black and red, it contrasted vividly with the familiar layers of shale and sandstone on either side.

It was like nothing they had seen before. They had, of course, been seeking such proof. But this surpassed their expectations, and promised to resolve the tantalizing scraps of evidence that had perplexed and divided scientists for so long, ever since exploration of the planet's past had begun.

Getting to the stratum was not easy. As luck would have it, it did not, here, extend down to river level, being cut off at the base by a huge natural dislocation in the rock strata, another inheritance from these mountains' violent past. The team had to climb halfway up that vertiginous ravine. But they had good balance, these explorers, their tails and sharp claws helping them scramble up the near-vertical rocky surfaces.

The hard material of the stratum formed a natural overhang, and, once they had climbed over that, they could explore in some comfort. There was an animated susurration as they communicated their discoveries. Here, an exposed rock surface with a regular, rectangular pattern, unlike any produced by normal geological processes; there, layers of angular pebbles with hard organic coatings. The remains of a long tubular structure, now oxidized, that had once been metallic. Parallel-sided shards of a white glassy substance. Another oxidized metal fragment, this time hinting at a complex internal structure: not a biological skeleton, but obviously manufactured.

There could now be no doubt. There had lived here, many millions of years ago, an ancient civilization, and one that could colonize on a grand scale: the stratum extended as far as their vision carried in the cliffs above. The explorers took samples from above and below that remarkable stratum, but the more experienced of them were convinced, already, of its deeper significance. It was at the same geological level as the traces of the ancient, catastrophic, environmental change that had, over years of their researches, emerged as an ever-clearer part of this planet's geological record.

So, the catastrophist school of thought was—well, perhaps not altogether vindicated, but at least they now had a basis in hard fact. There was now good reason to think that the ancient, planet-wide catastrophe had not been, as many had argued, a purely environmental crisis. Rather it had been associated with (or caused by?—the arguments would rumble on for many years yet, even as yet more astonishing evidence was to emerge) a major, intelligent yet transient civilization, many millions of years ago.

Of course, there had been signals in the rock strata that had hinted at such a thing. There were the major changes in animal and plant life, comparable to those yet more ancient biological convulsions that could be discerned even earlier in the planet's history. Strange chemical and isotopic signals were present in the rock strata. Isolated artefacts and fragmentary dwelling-structures had been uncovered. An ancient civilization? Not necessarily. For these appeared so suddenly in the geological record that they, it was argued, were more likely to represent earlier extraplanetary visitors, who left because of, or had been killed off by, the environmental vicissitudes of the time.

But now the doubters could be answered. This was a critical moment in the understanding of the planet's history, and the explorers knew it. The first undoubted evidence of a sophisticated civilization with the capacity to re-engineer part of the planet's surface. The silence that had accompanied this dawning realization was broken by a shrill whistle from one of the party. On one of the rock surfaces, a skull was showing.

Perspective

The purest of science fiction. The Earth, in a post-human future, many millions of years hence, being re-explored. By . . . who? Perhaps extra-terrestrial explorers or colonists, just as we now peer at images of rock strata sent back by the Mars landers. Or perhaps a new, home-grown intelligence: say, a newly evolved species of hyper-intelligent rodent. No matter. What would such explorers, of whatever ancestry, find of our own, long-vanished, human empire?

A frivolous question, perhaps. But perhaps not. It is hard, as humans, to get a proper perspective on the human race. We know that the Earth has a history that is long beyond human imagination, and that our own history is tiny by comparison. We know that we are animals, and yet we have transcended our natural environment to live in surroundings that, mostly, we have manufactured for ourselves. We know that this created environment is evolving at a speed that is vastly more rapid than the normal evolution of biological organisms or communities. We do not understand, quite, how our created environment and our activities interact with the natural environment, and we do not know what the long-term consequences will be.

Let us take one view. We are simply one species out of perhaps 30 million currently inhabiting the planet (reputable estimates range from some 5 million to over 100 million). We are briefly in the golden age of our power, our dominance. But we are destined to extinction also, just as the dinosaurs became extinct. The world will then go on as before. Once a geological age or two has passed, there will be nothing but the odd bone or gold ring to show that we were ever here.

In this scenario, comparison with the dinosaurs is apt. They were the top predators of their day, as our single species is now. But consider, also, the differences between us and the dinosaurs. The dinosaurs existed on this Earth for about a hundred million years, and included many species adapted to different environments. *Homo sapiens* is but one species, and has been around for less than a quarter of a million years, less than a tenth of an average species' longevity. Only in the last 200 years, since the Industrial Revolution, have humans had an unambiguously global impact. It is hard to compare the human and geological timescales. But take, say, a Greyhound bus to Flagstaff, Arizona. Make your way to the lip of the Grand Canyon, and gaze down. In that mile-deep chasm, the strata span 1.5 billion years. Measured on such a scale, our own species span would fit into a layer just three inches thick, while our industrial record should be confined to just one-hundredth of an inch.

Now, even though dinosaurs lived, collectively, over a vastly longer time span than have humans, their remains are strikingly rare. Despite intensive searches, only some few thousand skeletons that are anywhere near complete have been found, together with scattered footprints and occasional eggs. Significant discoveries become headline news. Why are dinosaurs rare? First, they were near the top of the food chain, and therefore relatively scarce. Secondly, they were dominantly terrestrial. On death, their bodies were generally exposed to the elements, and scattered and recycled by the myriad agents of scavenging and decay. Using such a comparison, the remains of our human empire should soon crumble away and decay, leaving scarcely a footprint on the sands of geological time. Our legacy would be as pitiful as that of Ozymandias' mighty kingdom in Shelley's poem, reduced to a shattered statue amid the boundless desert wastes.

Let us look from another angle. We are unquestionably the dominant life-form on this planet—numerous, intelligent, gregarious, self-aware. Our lives are dominated by contact with our own kind, to the extent that contact with the natural world for most of us is restricted to a walk in the park or a nature documentary on the television. Not only that, but the gulf between us and all other creatures is a chasm, an irreversible threshold in our (and the 'our' is emphatically possessive) planet's history. Nothing can stop us now: not war, nor fire, nor flood, nor plague. We will keep our grip forever, and go on to conquer the stars. The past four and a half billion years of the Earth's history has been nothing more than a preparation for our arrival. And now we have arrived. Nothing will ever be the same again. A million years from now, the Earth's surface will be covered with a finely engineered metal and plastic skin, and all of our needs—food, water, air, recreational wild beasts, and our own bodies (and souls, even)—will be computer-designed for undreamed-of comfort and pleasure.

But we can imagine other futures too, descending from grand confidence to utter pessimism. We are poisoning the planet, fouling our own earthly nest, causing ecological mayhem, producing an environmental *grande crise* which will not only cause our own extinction, but which will damage all present and future life on Earth beyond repair, and so put a full stop to the four-billion-year-long history of life on this planet.

'Nonsense!' others would cry. Life on Earth is in fine shape, ours included, and past extinctions have been on an altogether grander scale than anything we are capable of producing. We will carry on developing our industries and national economies to the common good, and Nature will carry on alongside us. For, volcanoes, surely, are vastly more polluting than chemical factories, and doomsayers have been predicting the end of the Earth ever since humans lived in caves.

Just how does one make sense out of these varied viewpoints?—they all seem individually so *plausible*. Just what is our true potential for immortality, and what might represent a true perspective of us as a single, newcomer species to this planet?

So here is one possible approach. Let us examine what our ultimate legacy is likely to be, the extent to which the human race and its actions are likely to be preserved within geological strata, and thus transported into the far future. This will be an acid test of our ultimate influence, our final footprint on the planet. Will it be deep, and permanent, or will it be quickly erased by wind and water once we are gone? We will confine ourselves to the geological footprint created by the human race up until the present, and into the near future, say the next century or so, in which some trends are relatively predictable. After that, most bets are likely to be off.

We will posit, as the forensic researchers into our legacy, extraterrestrial visitors from the galactic empire, finally arrived to this obscure outpost of the Milky Way. They will be given attributes of intelligence and inquisitiveness, which is a reasonable assumption. On Earth today, such qualities are possessed—to a perceptible degree—by crows, cats, and octopuses as well as humans.

The arguments put forward will not be affected by whether we become extinct over such a timescale, by any combination of plague, war, and famine. The longer the human species lasts, the deeper is likely to be the footprint. No special pleading need be invoked; one can simply apply normal geological principles to studying the preservation potential of humans and their handiwork. The estimates used will stay sober and conservative. Where different trajectories or options are possible, these will be spelled out.

Like other organisms, we may leave fossil evidence, in the form of our bodies and traces of our activity in the rock record, but this record will inherently be biased, and tricky to read. Human activities have also been changing the Earth's landscape, and so the way in which rocks themselves are being formed. They have also been changing the Earth's climate and biosphere. Will the effects of these changes be readable from the rock strata of the future? In our definition of 'future', a modest goal is set for the immortalization of traces of human activity: one hundred million years. A long time even for the most grandly far-sighted of empire-builders, but only just over two per cent of the Earth's current age. It is roughly the time span that separates us from the heyday of the dinosaurs.

Of course, not all fossils are immediately interpretable. That most aristocratic of dinosaurs, the *Iguanodon*, used to be reconstructed with a spike on the end of its nose, until it was realized—once more complete skeletons were found—that the spike was, in fact, its thumb. And there are a number of strangely shaped fossils which haven't yet been put into any broad category at all. So there will be occasional diversions, as one imagines our visitors of the future musing over the problematic history of this prodigal Earth. Will such alien palaeontologists find traces of *Homo sapiens*? Would they be able to reconstruct our bodies? How about our cultures, and how we interacted with the world around us? In other words, we will consider also *Homo sapiens* from the standpoint of a future palaeoecologist.

Such arguments about the preservation of physical remains of humans and of their activities are, of course, of little immediate practical value. Nevertheless, such an approach might offer a useful perspective on the current effects of human activity on Earth. For, there are big decisions ahead. If the effects of our collective activities are insignificant when set against the backcloth of natural environmental fluctuations, then there is little need to re-engineer, at great expense, our economies and our lifestyles so as to reduce our environmental impact. If, on the other hand, we are responsible for a perturbation of the Earth's baseline geological processes that will be detectable into the far future, then any efforts we make now to restore equilibrium would simply represent sensible life insurance policies for us all.

In this attempt to reconstruct (or pre-construct, perhaps) the way in which our future explorers might put together the geologically brief history of our species, one needs—as *they* will need—to understand the planet that could incubate such a species, and then preserve evidence of its existence. This means considering the complicated and rather wonderful workings of the Earth machine that will control our future preservation. These are, of course, exactly the same processes that have produced all the Earth's geological strata and the fossils that they now enclose. There is little reason to believe that they will work any differently in the future.

One must emphasize the patience that will be needed here, by our future explorers. Our remains will not turn up, Dr. Who-style, in a cave, brittle and shrouded with cobwebs. They will need to be uncovered by persistence, by logic, by following trails of clues—and of red herrings. It will be vital for these explorers to work out the geological blueprints for this planet, for some part of us will have become geology. What remains of us, though, will inhabit a brief geological instant that will seem lost among many millions of other such instants that have succeeded one another in the long history of the Earth.

On the other hand, there might be reason to suppose that the strata of the Human Period may turn out to be quite distinctive, once the geological signals are decoded and converted into history. They may be as recognizable, perhaps, as the strata, some 540 million years old, that mark the first extraordinary explosion of multicellular life on this planet; or as distinctive as the thin stratal interval which contains the traces of the meteorite impact which, 65 million years ago, was coincident with the abrupt termination of the dinosaurs' long reign.

One can finish this introduction on a level, finally, that we can say is more human—if not downright personal. If *you* desire immortality for some aspect of your own personal sojourn on Earth, then these pages might contain some more or less soundly based practical advice on how you might increase your chances of carrying a final message, that of your own brief existence, into the next geological era. If you wish, then, to adorn some museum of the far future, read on.

100 Million Years AD

We have sighted a most remarkable planet. Remarkable enough, I think, for the news, when it is eventually relayed, to cause great interest and excitement at home. We are still a long way out, but it is clear that this planet has an extraordinary and seemingly quite unstable surface chemistry. Our sensors have detected not only much free water at the surface, but also—and this has surprised us—free oxygen, and in considerable quantities.

FUTURE EARTH: FIRST SIGHTING

A storyteller arrives, one hundred million years from now, to tell the tale of the human species. It is an interval that will add a couple of per cent to the age of the Earth and a little under one per cent to the age of the Universe. Geologically, it is the near future. Cosmologically, we are almost there.

There will be an Earth, that which we now call our own. On it there will be, very probably but not quite certainly, oceans of liquid water, an oxygen-rich atmosphere, and an abundance of complex, multicellular life.

The Earth is abnormal, and that will draw any interstellar travellers in. The spaceship's sensors—a simple spectroscope will suffice here—will immediately register the highly reactive surface chemistry that is out of any sort of normal equilibrium. An oxygen-rich atmosphere is not normal. Even from a distance of many millions of miles, this will be a planet that is obviously alive.

Closer up, the living skin on the planet, regulator of that planetary surface chemistry, can begin to be glimpsed, as the green wavelengths that mingle with the blue of the oceans of liquid water and the brown of the rock surfaces. Our future visitors would not yet be aware of chlorophyll, but that unexpected signal shining through in the light spectrum would certainly arouse their curiosity.

Rock, oceans... and green stuff. The geography of the Earth, to our own human and contemporary eyes, would look oddly familiar, but distorted: as though remodelled by Salvador Dali. Familiar landmasses will be displaced. But where to? Unfortunately, we cannot predict where the Earth's continents will be in one hundred million years' time. Will the Atlantic Ocean continue to widen, and the Pacific Ocean shrink? Will the East African Rift expand into an ocean? Will the continents aggregate into supercontinents, as has happened in the past?

Long-term tectonic forecasts, like long-range weather forecasts, are subject to such uncertainties that detailed prediction becomes useless; there are simply too many possible alternative futures. Our planet's physiography will simply be different, one hundred million years from now, though with elements we would find partly familiar, rearranged as though by the hand of some gigantic and playful child.

If the future physiognomy of the Earth is uncertain, then so, too, is its temperature setting. For global climate depends, to a large extent, on global geography. If future geography is not previsible, then neither is future climate. One of the reasons we live in an age of ice is that the current pattern of oceans and continents does not allow ocean currents easily to spread the warmth of the tropics towards the gigantic, well-insulated storehouse of cold that is Antarctica.

One hundred million years ago, in the Cretaceous Period, the patterns of continents and oceans was different, and tropical warmth spread so far north and south that there was little or no (opinions differ between 'little' and 'no') ice at either pole. Sea level, in consequence, was at least 70 metres higher than it is today, and far more of the continents were submerged.

So which future to choose—global hothouse or icehouse? One can simply use historical precedent. Hothouse times were more common than icehouse times in the Earth's geological past. One can also invoke astrophysical inevitability: the sun will be very slightly hotter, and so will have pushed the Earth just that little bit closer to the point at which the oceans will boil away and all life will cease. So, with or without human help, the Earth would more likely than not revert to its usual mode, with a warm climate—as enjoyed by the dinosaurs—and without significant ice caps.

So . . . a warm world, with tepid rather than chilly ocean waters; a bluer world than today, with nearer three-quarters than two-thirds of its surface underwater. A green world still, on those landmasses that emerge above the global ocean. If our far-future visitors have a sense of wonder and an appreciation of beauty, they will be entranced. Time to make a landing.

FUTURE EARTH: CLOSE-UP

The living world they will encounter will astonish them with its variety, though with new and—to us—imprevisible dynasties of plants and animals. For who, landing on Earth one hundred million years ago, in the time of the dinosaurs, could have foreseen the present dominance of mammals and flowering plants?

One might imagine, perhaps, a diversity of rodents derived from our present-day rats, for these have been fellow-travellers of humans all around the world, and have proved durable and enterprising colonists in their own right. Their descendants may be of various shapes and sizes: some smaller than shrews, and others the size of elephants, roaming the grasslands; yet others are swift and strong and deadly as leopards. We might include among them—for curiosity's sake and to keep our options open—a species or two of large naked rodent, living in caves, shaping rocks as primitive tools and wearing the skins of other mammals that they have killed and eaten.

In the oceans, we might envisage seal-like rodents and, hunting them, larger, ferocious killer rodents, sleek and streamlined as the dolphins of today and the ichthyosaurs of yesteryear. Among them, fish—both bony fish and the cartilaginous sharks—might still be prominent in the oceanic ecosystem. Perhaps, though, as populations of the latter are now plummeting—vertiginously so in the case of the bigger sharks, because of human over-fishing—this might give a chance to other creatures, perhaps the squids, to expand further into this niche, perhaps even to take over dominance within it.

Our explorers will have their hands full, for some quite considerable time, simply in examining and cataloguing this plethora of living organisms on both land and sea. From their perspective, though, they may be as much interested in the overall design and fundamental controls on this world and its ecosystem, as in the details of its almost infinite variety.

Thus, they will see that the base of this immense diversity is the plant life, capturing a portion of the light from the sun and reflecting the unused portion of the spectrum back into space (aha . . . they may say as they work that out—so that's where the green colour of this planet comes from). Then using the energy from that captured light to split simple molecules of water and carbon dioxide, recombining them into complex organic molecules which they use to build themselves, releasing oxygen into the atmosphere as a by-product. And, by further sophisticated chemical sleight-of-hand, these plants extract the stored energy in some of their newly synthesized molecules of carbohydrate by reacting them with oxygen, converting them back to carbon dioxide and water, and using the energy derived from this oxidation, that we call respiration, to drive all of their internal biomolecular mechanics.

Then, there are a whole array of organisms of different types on the Earth—the animals—that lack that ability to trap sunlight and to synthesize the stuff of their own bodies from simple raw materials. These exploit the plants as a source of energy and materials, mostly by the simple expedient of eating them. Some of the animals even bypass the plants altogether, to exploit richer material stores, the bodies of some of their fellow animals that they have to catch and kill for this purpose. We assume that our explorers, being familiar with life, will be familiar with the intricate ecological webs along which matter and energy flow, that this life creates from its millions of species and trillions of individuals. We may be mistaken in this assumption, for we are ignorant of other possible designs that living systems may take, on other planets around other stars.

No matter how familiar or how strange our explorers will find the underlying dynamics of the living system of this particular Earth, they will—we will assume—be avid to understand exactly how it functions and how it originated. Thus, the physical and chemical context of the life of this Earth will figure largely in their investigations. Further, to understand the pattern of life that they will find and to illuminate its relation to the solid planet itself, they will need to understand the history of the living organisms and of the planet that they inhabit.

That history, they will soon realize, covers the surface of the Earth as it covers no other body in this solar system. Litters it, in fact. For the Earth is a treasure-house of strata, and it is to the strata that they must soon begin to turn their attention.

The Strata Machine

Such an abundance of life! Our biologists have work to occupy them here until eternity. So many varieties, and yet linked to each other, with the same molecular blueprint. And so mineralized!—many of the life-forms here make wonderfully intricate crystalline structures to live in, or to arrange their living tissues around. Where has this life come from? How has it arisen? There may be clues to this history in the layers of sedimentary debris that abound on this surface. In scale, they far exceed anything on the neighbouring planets. We need to explore further.

History is bunk—or so Henry Ford is reputed to have said. Folk memory, though, simplifies recorded statements. What Henry Ford actually told the *Chicago Tribune* was 'History is more or less bunk. It's tradition. We don't want tradition. We want to live in the present, and the only tradition that is worth a tinker's damn is the history that we make today.' So folk memory, in this case, did pretty well reflect the kernel of his views. Henry Ford also said that 'Exercise is bunk. If you are healthy, you don't need it; if you are sick, you shouldn't take it.' Henry Ford was a very powerful, very rich man of strongly expressed views. And he was quite wrong on both counts.

Not having known Henry Ford, interplanetary explorers may have their own view of history. As, perhaps, an indispensable means of understanding the present and of predicting the future. As a way of deducing how the various phenomena—physical, chemical, and biological—on any planet operate. And as a means of avoiding the kind of mistake—such as resource exhaustion or intra-species war—that could terminate the ambitions of any promising and newly emerged intelligent life-form.

On Earth, and everywhere else, things are as they are because they have developed that way. The history of that development must be worked out from tangible evidence: chiefly the objects and traces of past events and processes preserved on this planet itself.

The surface of the Earth is no place to preserve deep history. This is in spite of—and in large part because of—the many events that have taken place on it. The surface of the future Earth, one hundred million years from now, will not have preserved evidence of contemporary human activity. One can be quite categorical about this. Whatever arrangement of oceans and continents, or whatever state of cool or warmth will exist then, the Earth's surface will have been wiped clean of human traces.

For the Earth is active. It is not just an inert mass of rock, an enormous sphere of silicates and metals to be mined by its freight of organisms, much as caterpillars chew through leaves. Nor will it be inert, a hundred million years into the future. It is a dynamic system, powered from inside by the heat generated from the radioactivity within its interior. The internal radioactivity of the Earth is in truth no greater than that of surface rocks, but because of the huge bulk of this planet it escapes only slowly, and hence the temperatures of the Earth's interior have remained at the point at which rocks begin to melt.

But escape this heat must, and future visitors may be puzzled as to just how this trick is carried out on a relatively large planet, that produces very large amounts of heat. Obvious symptoms of terrestrial heat release will be present, as volcanoes will erupt more or less as frequently then as they do now. But volcanoes are two a penny, even in this small solar system of ours. Venus possesses many. Mars has the largest in this solar system, the mighty Olympus Mons. Io, that satellite of Jupiter, possesses the most active volcanoes, constantly pouring plumes of sulphur out into its thin atmosphere.

There will remain a bigger and considerably more subtle question. How can the internal dynamics of the Earth operate to produce such an interestingly variegated surface, where landmasses rise high above that ocean?

For those landmasses are under continual attack as the Earth's second energy source—the radiation continuously received from the sun—drives the motion in the Earth's fluid envelope of air and water to create powerful weather systems. The resultant wind and rain and waves, combined with the corrosive effects of that ultimate solvent, water, inexorably destroy the rocks of the land surface. Mountainous landscapes are, over millions of years of time, eroded down to sea level. But the Earth has been in existence for billions, not millions, of years. To maintain mountain-bearing landmasses necessitates their continual renewal, with enormous rock masses being pushed high up above the surface of the ocean as fast as they are being worn down.

Thus, one hundred million years from now, nothing will be left of our contemporary human empire at the Earth's surface. Our planet is too active, its surface too energetic, too abrasive, too corrosive, to allow even (say) the Egyptian Pyramids to exist for even a hundredth of that time. Leave a building carved out of solid diamond—were it even to be as big as the Ritz—exposed to the elements for that long and it would be worn away quite inexorably. There is a splendid Arthur C. Clarke story, *Against the Fall of Night*, about an engineered, enclosed human colony, a huge tower, surviving in the arid wastes of an ecologically devastated world for hundreds of millions of years. It is a marvellous plot, but a quite implausible scenario. The weather would shred the tower, or undercut it, or bury it, in a tenth of the time.

So there will be no corroded cities amid the jungle that will, then, cover most of the land surface, no skyscraper remains akin to some future Angkor Wat for future archaeologists to pore over. Structures such as those might survive at the surface for thousands of years, but not for many millions. A sense of the historical possibilities of this planet, though, might well strike our future explorers. The superabundance, on this Earth, of strata.

Strata! Just the layering of rocks, a simple physical property. Nevertheless it captures within itself not just the arrow of time, but almost infinite possibilities for encapsulating stories of past landscapes (and—much more often, on the Earth—of vanished submarinescapes); and of the organisms that lived and died on them, through almost unimaginable reaches of deep time.

Even the most seemingly inert of the rocky bodies of this planetary system preserves strata. The single moon that orbits around this planet has been effectively frozen for the best part of four billion years. Yet it too possesses strata, of a sort. The debris flung from the sites of meteorite impact, for instance, form aprons of rubble surrounding the craters extending, in the cases of the largest impacts, up to thousands of kilometres away from the crater. The layers from the countless impact sites on this moon will, here and there, overlap each other, criss-crossing to form the beginnings of a lunar layer-cake of rubble strata.

Those layers preserve their own history. The manner in which an impact-layer piled up—perhaps thicker on one side of the crater than the other—can preserve the story of the speed and flight direction of the projectile, just before it made contact with the Moon's surface. The size of the fragments, the degree to which they had been flash-melted and their state on reaching the surface (still-molten, say, or already solidified) will tell further stories about their almost instantaneous birth as the crater was excavated, and of their brief flight over the lunar surface. The distribution of these pulverized rock fragments may, indeed, give clues as to whether any atmosphere existed above the Moon over 4 billion years ago, or whether that satellite has always been airless, for the dynamics of flight of impact-driven particles through a vacuum are different from those through an atmosphere.

Then, there were the lavas that later, and briefly, welled up out of the Moon's interior, when that interior was still hot. They effectively form layers, too, piled up one above the other, infilling the depressions of the early Moon. These frozen seas of lava strata, visible as dark patches from the parent planet, were to become seas of liquid water in the human imagination for many generations. Humans ultimately reached that moon in spacecraft, to discover a moon that is essentially anhydrous, drier than the driest and hottest parts of the Sahara desert or the Empty Quarter of Arabia. But those lavas, too, tell their own story, of the chemistry of the interior from which they came, during their eruption two billion years ago and more.

There are better-developed strata on Mars. The planet itself will likely have changed little, one hundred million years from now. It may be more or less frosted over by its polar caps of ice and frozen carbon dioxide, but its fundamental landscape will be little altered, bar a few extra meteorite craters. Olympus Mons will still tower higher than any other volcano in the solar system, while that remarkable planetary-scale scar, the canyon of the Valles Marinaris, will be just as deep and precipitous. These giant landscapes of solar system topography will be slightly more eroded than they are now, as the winds of the thin carbon dioxide atmosphere blow abrasive dust across the landscape, and drive Martian sand into dunes that look remarkably like those of the deserts of the Earth. These modern and future sand-dune strata, though, are just a surface skin that lie, here and there, on a planet that has been freeze-dried for more than three billion years.

These wind-driven sand layers lie atop thicker, more ancient strata that tell (or seem to tell, to current human eyes) of a more active and a wetter ancient history, a history of extensive planetary surface water that was ending as the history of the Earth's oceans was in its infancy. These strata, given the almost-fossilized state of the Martian surface, continue to be visible for billions of years into the future. There is a striking, widely reproduced image in our textbooks of... well, not exactly a river channel, but rather of the strata deposited by a river channel as it meandered more than three billion years ago across the surface of the planet, leaving a trail of bars of river sediment in its wake. From those strata, even without being able to land beside them, measure them, hammer them, it is possible to work out which direction that Martian river was flowing in, and how it changed its course in its brief history. It would be nice to have a closer look at this palaeo-river, to examine the grains and pebbles that were carried by it, to measure the length and height of the ripples and dunes that formed on its banks. How long did that Martian river flow for, for example?—and how fast did the water flow?

There is, though, a history that is much closer to us, that is more eloquent and extravagant by far, and more scientifically tempting. The Earth, by comparison with these neighbouring planetary bodies, is a treasury of strata, a gigantic machine for producing strata that contain within themselves countless narrative possibilities of the histories of former oceans and rivers, of lakes and shorelines and arid deserts, of anywhere, in fact, where sediment can accumulate at the Earth's surface. Moreover, it is one that continues to function today, and it will be functioning one hundred million years from now, much as it has functioned for billions of years. It has provided a history machine, a dazzling capsule of planetary memories that is without compare in this solar system.

A planetary explorer on the Earth could not fail to notice this, even though many of the land surfaces that they will explore will be thickly vegetated, this living cover draping and concealing the underlying geology. Today, when one is strolling through gentle rolling countryside, through meadows and fields, it is hard to imagine that one is walking over rock strata, often not much more than a few tens of centimetres beneath one's feet, beneath a continuous cover of soil and vegetation. Amid the deserts or the mountains of the Earth, the underlying geological bones of the landscape emerge much more clearly. Here, the topographic ridges formed by the hard rock strata stand out clearly, the rocks themselves being exposed at the surface as crags and cliffs.

These strata patterns betray their nature, even when gazed at from afar. Think of the parallel cliffs and benches of the Grand Canyon, or the layered cliffs of the Dorset and Yorkshire coasts in Britain. Some of these compare nicely with those seen on other planetary surfaces. There is a fine wave-cut platform north of Scarborough, on the Yorkshire coast, that can be looked on from the top of the present-day cliff, where the strata take the form of a meandering river channel, dried out and buried many millions of years ago, and now re-excavated by the waves of the North Sea. It is a close parallel of that fossilized river from Mars. And there are other examples: there are series of cliffs in Utah, equally old, even more scenically splendid, that bear distinctive diagonally slanted strata. These crosswise strata represent the steep surfaces of fossilized windblown dunes that were once blown across the ancient North American continent. They are like the surfaces of the dunes that are now being created by the desert winds in the Sahara—and like those now migrating across the surface of Mars.

They are eloquent of past geological environments and processes, these strata now exposed on the Earth's surface. Yet, they are quite impermanent geologically. The wind and waves and rain wear them away, releasing loose sediment into streams and rivers and on to beaches. The sediment particles, on this journey, are sorted and re-shaped by the ceaseless motion of the Earth's gaseous and watery envelopes and become, in turn, future strata in the making.

Such strata are not just landscapes caught in time but those landscapes preserved *through* time. Where sediment particles accumulate on shorelines, say, they might ultimately be preserved as fossilized beaches. Not just a beach surface, the two-dimensional shape of that gently sloping sea-facing apron of sand and rounded pebbles. Dig a little deeper, and the beach becomes a layered stratal unit, with a thickness of perhaps a few metres. Within it, the gently inclined layers represent thousands of individual former beach surfaces, superimposed within that layer of rock. These tell a story of the history of that beach through many individual beach-forming events, tracking, say, storms and fair-weather episodes and the action of the tides. The interpretation of strata therefore does not give a series of tableaux, of ancient landscapes frozen in time. Rather, it is cinematic, showing those landscapes changing and evolving as they reacted to the forces that shape them. There are often breaks in the film, true, but the basis on which the forensic examination of rock strata rests is one of continuity, not stasis.

On Earth, now, it seems obvious to us to link layers of hard rock with the sand and silt that is washed by waves and currents down rivers and along beaches. Yet, it is a step that took human civilization thousands of years to make, even as our ancestors were hewing sandstone and limestone to build with, and digging deep into the Earth for flint and salt, iron and copper. For the conversion of sediment into rock typically takes place in inaccessible regions of the planet (deep underground) and takes great lengths of time. This is a process far removed from the short life and geographical range of a pre-industrial human, even if they had the leisure and the wish to consider such a possibility.

The production of sedimentary rock from primordial Earth-rock is the process by which the Earth has acquired a tangible history. More: this history includes a detailed record of the evolution of life, and, ultimately, of intelligent beings, a story more intricate by far than the recording of physical and chemical phenomena. On a planetary scale that makes the Earth a quite singular object. Its production line for strata is distinctive and elaborate. Some aspects of it might be quite familiar to our future visitors, because they will be common to many rocky planets—the making and fossilization of sand dunes and river channels, for instance. Others will appear to be unique: there are no other equivalents in this solar system of, say, a coral reef limestone.

What, though, is our starting point, our primordial Earth-rock? A reasonable definition here might be to call it the rock that crystallized out from a molten state. Early in its history, the Earth would likely have had something approximating to a magma ocean, and as this slowly cooled, the molten rock crystallized to form a crust of solid material of igneous rock. The Earth is still, more than four billion years later, hot enough to be partially molten inside, heated from within by its remaining radioactive content, and this magma can gather into subterranean chambers and, being less dense than the adjacent solid rock, it can then work its way towards the surface. There, it can either break through in volcanic eruptions or slowly cool beneath ground level to form masses of rock such as granite and gabbro.

Such igneous rocks were born at temperatures of several hundred degrees centigrade, and often at the high pressures of the Earth's interior. The minerals out of which they are made grew in equilibrium with such conditions. Taken to the surface, and once cooled to temperatures of only tens of degrees centigrade, their component silicate minerals are no longer in equilibrium with their surroundings. Their molecular structure is stressed, no longer in its optimum arrangement. At these lower temperatures, the molecular structures become poised to undergo a rearrangement to suit their new conditions. Yet, they can remain poised, and the minerals can remain in a metastable state virtually forever, if nothing catalyses their breakdown. In this way the minerals of the igneous rocks on the Moon are as fresh as when, a few billion years ago, they crystallized.

This is because the Moon is bone-dry. Just add liquid water, though, particularly the weakly acidic rainwater widely available at the Earth's surface, and those silicate minerals, forged at high temperatures, disintegrate in this new corrosive environment. The components reassemble themselves into minerals that are stable in their new, cooler and wetter surroundings. Only a few of the primary igneous minerals resist the surface rotting. The most common of these, quartz, is simply released as physical particles while other minerals decay around them. Quartz grains, hence, form the great bulk of the sand of beaches and river channels and desert dunes.

Most of the original igneous minerals—olivines, pyroxenes, amphiboles, micas, feldspars—unravel, and are stripped down. Their molecular structures break down, transform into new structures, into tiny flakes of new mineral. Phoenix-like, the clays arise from this destruction.

Clay minerals are tiny, but their size belies their complexity. Looked at with the most powerful of microscopes, they resemble complex Meccano multi-storey buildings, the struts and girders being made of silicon, oxygen, and aluminium ions stripped down during the disintegration of the igneous minerals. They may form amid the molecular ruins of the rotted igneous minerals, or simply grow from solution on the surfaces of pores and fractures in the rock. The floating ions snap into place, as in a chemical garden, to form microscopic flake-like crystals, a hundredth of a millimetre across or less. Clays are the main ingredient of the Earth's signature sediment: mud.

Mud gets a bad press in polite human society, but the hippopotami in the Flanders and Swann song got it pretty well right. It is glorious stuff, and it happens to be utterly indispensable to the Earth as we know it. Together with its compacted and indurated offspring, mudrock, it forms a large part of the solid surface of this planet. A newly arrived visitor to this planet would register its abundance with some considerable attention.

Mudrocks make up the majority of all sedimentary strata, and most surface soil and sediment includes at least some mud. In any attempts at comparative planetology, some may see the Earth as the living planet (alone in this respect—or as near so as makes little difference—in the solar system); or the green planet, for its thick carpet of terrestrial vegetation; or the blue planet, for its deep oceans. But one might equally denote this planet as the muddy planet, for it is the only one to be encased in a thick shell of mud and mudrock, being literally enveloped in its own decay products. This planet is deeply rotted. There is a shell of chemical alteration products around it so thick that it can be difficult to find primordial rock.

Mud is indispensable to the functioning of the Earth's life support systems, because of the importance of the numberless clay particles to the Earth's geochemical cycles. It seems also to have been indispensable to the origin of life on this planet, for the reactions by which amino acids react together to form more complex organic structures proceed far more quickly in the presence of clay minerals. Clay minerals are not unique to the Earth—they have been detected, in small quantities, in Martian strata, having been formed in the brief early warmer and wetter phase of that planet. The Earth, though, is plastered in mud, like a child emerging from a football pitch on a rainy day.

Mud, though, is more than clay. Complex as the clay minerals are in structure, they pale (literally) beside the ferociously complex remains of life and death that abound in most muds, that form a rich, dark, and—quite frankly—smelly mulch of decomposing organic matter in this sediment, feasted upon by countless bacteria.

Mud is therefore one of the world's great carbon stores, a planet-spanning communal tomb for the composted remains of many generations of living organisms. Buried, heated, and compressed, the complex hydrocarbons break down into simpler hydrocarbons that migrate underground, in places accumulating as underground reservoirs of oil and gas. Or, these fluid hydrocarbons might simply leak back to the surface, to be absorbed by and reborn into new generations of living organisms. These mud-derived carbon stores also act as a crucial control on climate, not least when occasionally exploited by energy-hungry civilizations.

Other decay products of primordial rocks are released, as charged ions, into solution: sodium, potassium, calcium, chlorides, sulphates, carbonates. These have, over time, accumulated in the oceans, rendering them salty. The oceans, indeed, have absorbed so much of some of these decay products that they can scarcely hold any more. Surface ocean waters are saturated, for instance, with respect to the combination of calcium and carbonate ions. These link easily, either by simple chemical means or via the intervention of plants and animals, to form the mineral calcium carbonate, the prime ingredient of limestone.

Limestone constitutes another gigantic carbon store on Earth, greater even than that within the mudrocks. Indeed, the Earth's limestone strata have, locked away within them, something of the order of one hundred atmospheres' worth of carbon dioxide. Our future explorers will quickly realize the significance of this as a mechanism for the long-term regulation of planetary temperature: this rock-bound carbon store is roughly the same size as that present as carbon dioxide gas in the atmosphere of Venus. It is an amount that traps enough radiation at the surface of that singular planet to maintain its temperature at a furnace-like 400 degrees centigrade, a temperature that has boiled all water away from both its surface and atmosphere.

Mars, curiously, has very little carbon dioxide either in its atmosphere (though what little atmosphere it has is mostly of that gas) or, surprisingly, locked away as mineral in its surface strata. Either Mars had little carbon to begin with (which seems unlikely) or its carbon dioxide was lost to space, slowly 'sputtered' from its atmosphere by solar radiation, a process made easier by the weak gravitational field of that small planet.

It is becoming ever more clear that strata do not simply form a rocky shell to the planet, of scenic and historical interest to curious passers-by. Rather, they have played—and still play—a crucial role in maintaining and regulating stable conditions for life on this planet. Life has not been passive in this regard. It has produced some of the Earth's strata—many limestones and all coals, say—and its remains, such as petroleum, have suffused others. And it has left fossils.

The strata of the Earth are graveyards, the burial places of relics from which long-dead organisms can be recreated. Visiting aliens will certainly be aware, on general principles, that things of this sort are possible. Something must happen to the body of any living being, after death. Being a direct source of the building blocks of life, any dead organism would be typically recycled in any reasonably stable, self-sustaining environment, via the combined action of a variety of browsers, predators, scavengers, and, finally, by the ubiquitous microbes. The great tropical forests, one hundred million years from now, likely to have recovered their former extent and glory, are superb examples of this. That abundant biomass is held within living matter. On death, the tissues are broken down to their constituent atoms and molecules and these then help build new organisms, as life is continually cycled from generation to generation. There is little trace of any of this biological wealth in the poor, thin soils beneath.

But, the earthly remains of dead animals and plants are not always and not everywhere recycled, particularly those parts of the organisms that are mineralized and hence decay-resistant, such as the shells of molluscs and the bones of vertebrates. The biochemical trick of skeleton-building was acquired relatively late in the three-billion-year history of Earthly life: in the last half-billion years only. It hinged on a particular grade of biological development, but also on the surface chemistry of the Earth, with its readily available raw materials of carbonate and phosphate. And, it also reflects the emergence of a biological or ecological need for such structures, either for defence or attack, or for a skeleton to hang soft tissues on. Once one group of organisms has discovered this trick, there will be a greater need, translated into intense selective pressure, for other groups to follow, if they are not to be left behind in the evolutionary arms race.

Prolific skeleton-building and shell-growth may be a peculiarity of Earth. It is not clear that life on other planets, developing different bodyplans on surfaces of different chemical compositions, would necessarily grow hard skeletons. There may be planets colonized solely by soft-bodied organisms, but it is far harder to envisage one that has not seen the emergence of bacteria, those great recyclers of dead (or even living) tissue. On such planets, there would be little scope for the smuggling of biological relics across deep time into any geologically distant future, and palaeontology as a science could hardly get started.

On an Earth colonized only by soft-bodied animals and plants, the palaeontological record would be meagre. There would be scattered indistinct impressions on stratal surfaces, in the way that jellyfish are, exceedingly rarely, found fossilized. Each example is a museum piece. Even a thin shell is no guarantee of preservation. Krill, for example, are lightly armoured crustaceans that are so abundant that they support a good deal of the Earth's whale populations. Yet, they are almost never found fossilized.

But many organisms have more robust skeletons. Once one begins to examine rock exposures closely, it only takes a little exploration to discover that the Earth's strata are a treasure-trove of the fossilized remains of long-dead animals and plants. Fossils are not hard to find, once you are looking for them and have some inkling of what they represent—that is, once you have made the link between living present and ancient past.

Limestones are full of—and indeed many are virtually made up of—the remains of corals and other skeleton-building creatures. A good part of the English Pennines, for instance, is made up of limestones that are often clearly made of a mass of fragmented and packed stems of calcareous sea-lilies, looking like so many broken pipe-stems. The fine-grained nature and chemical reactivity of mudrocks, too, often make them an exquisite medium in which the remains of living creatures are petrified. Even the relatively large quartz grains of sandstones may retain fossil imprints.

Fossils are everywhere, almost, and it seems surprising that our own human species took so long to take those curiously shaped objects that came out of rocks and connect them with once-living creatures of an ancient past. Indeed, the link was not explicitly made in Europe until the seventeenth century, by the Danish scientist and physician Neils Stensen, usually termed Steno, though earlier scientists such as Leonardo da Vinci, Xenophon, Herodotus, and Pythagoras had glimpsed the true meaning of fossils. In China, meanwhile, fossils had been explicitly identified as the remains of living organisms from the first century BC, with the recognition of 'stone fishes' and petrified 'swallows' (the latter were something of a misidentification, being based on the elegantly wing-shaped brachiopod *Spirifer*). The early Chinese insights were to be ignored by the rest of the world for almost two millennia.

This realization (in the West, at least) seems astonishingly late given just how similar some fossil shells are to their modern-day counterparts. It partly reflects the fact that many fossils are not immediately recognizable as the remains of long-dead creatures. Fossil graptolites, for instance, may resemble geometrical mineral growths, while genuinely inorganic mineral patterns may resemble petrified ferns or mosses. More generally, this late recognition of fossils and of what they represent probably gives some inkling of just how mysterious, how baffling, the world and its infinite variety of marvels seemed to our ancestors.

Any newly evolved, home-grown terrestrial species would, thus, in trying to understand the phenomenon of fossilization, need slowly and haltingly to retrace the steps taken by Steno, and then by James Hutton and Baron Cuvier and Charles Darwin and their successors. And they would—must—do so from an Earth-centric point of view. Interstellar visitors, though, will see the Earth in the contexts of other planets flown past, probed, visited, explored.

Broad patterns will emerge. There will be a general resemblance between the fossilized remains recovered from the rock strata and the myriad organisms that live on Earth. Thus, land animals have legs rather than wheels, and those that fly have wings rather than rotor blades. There will be abundant small shells made of calcium carbonate in the form of various flat to helical spirals or of paired valves. There will be carbonized impressions of leaves and twigs. These are generalities, of limited application, except to say that the living and dead are connected in some way. They invite further questions, such as whether particular forms impressed within strata have exact counterparts among living animals or plants (and, conversely, whether particular fossils have no living counterparts). If a fossilized organism has no exact living counterpart, one can begin to say that organisms have changed between the deep geological past and the present. This leads on to questions of how such biological change may have taken place through time. The explorers can begin to recreate the story of biological evolution, a story that may incorporate truly universal phenomena (Darwin's concept of natural selection, that should apply on any planet) and more local mechanisms (the particular workings of the Earthly genetic code and of terrestrial biochemical pathways). By whatever means, these strange petrifactions of past life can be used, now and in the future, to interpret biological process.

It will not be an easy task. Palaeontology is not a science for the impatient. It is necessary, first, to understand the living before interrogating the dead. Alien scientists cataloguing the Earth's living organisms will need to cope with their unfamiliarity of morphology and physiology and ecology, their very particularly Earthly distribution into sexes, and with their sheer number and variety.

Even we, as home-grown scientists with a close interest (mostly exploitative or defensive, admittedly) in the plants and animals that share our planet, do not know to within an order of magnitude how many species now inhabit the Earth. Indeed, our interplanetary visitors will need to come to terms with the concept of 'species', which is something that Darwin struggled with, and which we have yet to understand perfectly. But the number of forms, or varieties, or inter-breeding entities of organism (the 'breeding' business might take some understanding for an extraterrestrial, too) currently is likely somewhere between ten million and a hundred million. A large proportion of those are single-celled and microscopic and lack nuclei, the bacteria and archaea and viruses. These may well be regarded as the most important inhabitants of this planet (for good reasons, for without them many Earth surface processes would not function). Many species, even of the larger and more flamboyant multicellular organisms, look very much like each other: they are almost indistinguishable to human senses, let alone to the senses possessed by visitors from afar.

But if we assume that these future explorers have a level of technological sophistication consistent with interstellar travel, they may be able to employ detailed biochemical analysis to help classify the molecular identities of different organisms. And then, to identify them by their unique DNA signature—once DNA had been recognized as the key molecule that allows the propagation of life on this planet. These biochemical signatures can then be used to establish groups of related organisms and to determine the connections between them. Part, at least, of the whole business can be automated, much as genetic typing is carried out today. This would speed up the exploration of Earthly life considerably.

No such luck for the palaeontologist. Here one must deal with a variety of fragmentary shapes, impressions, mineral casts, and moulds, often squeezed or crumpled by the weight of overlying strata or by earth movements. Sometimes the stuff of the original animal or plant may be present, mostly it is not. If it is present, it is almost always altered in some way. The whole organism is never preserved (rarely, one finds almost complete specimens). Often the portion that is preserved may be as small as the pollen of a plant or the tooth of a shark. Simply finding the fossils may involve sifting through tons of rock and examining acreages of stratal surfaces. It is slow work, and resistant to automation. The modern human palaeontologist, despite having a variety of fine gleaming new machines to play with, such as CAT scanners and electron microscopes, essentially does palaeontology in the same way as did our Victorian forebears, only with more care and attention to detail, for the distinctions to draw have become finer.

Recognition of what the fossil might be is simply a comparison of shape with shape. It is the ability to put together n-dimensional jigsaws that matters: those where the shapes never quite fit together perfectly. And a selective retentive memory is essential, that can remember shapes from one puzzle to another, even if these puzzles are tackled a decade or more apart. Computers have, thus far, not been very good at this, while human minds and eyes, helped by stored information in the form of large, much-thumbed palaeontological monographs, have been remarkably effective.

An interplanetary scientist, coming fresh into this world of wonders, can start by being a polymath, a generalist, much as were the geologists of the eighteenth and nineteenth centuries, such as Alexander Humboldt and Carl Ferdinand Roemer and Charles Darwin himself. The task, though, grows greater year by year as more fossils are described and more monographs are written. Then, unless one were endowed with unlimited time and formidable computing power, there would be a need to specialize, to work only on this group or that one. Even more than with the study of living organisms, the task is never-ending.

Classifying and cataloguing shapes imprinted upon strata—what then? Well, they, and the strata, need putting into order. The principles of this ordering are not difficult and can be repeated on any rocky planet. The fruits of the research on Earth, though, are immeasurably greater than can be achieved on its neighbouring planets. For a new colonizing and analytical intelligence, this is probably most easily achieved starting with the present, and then working backwards on the long journey through ever-further reaches of deep time.

Modern beaches and sea floors are often littered with the shells of dead molluscs. Dig a little below the surface at these places, and similar shells are typically also found preserved in the shallowly buried layers of sediment. There may be fewer of them, for those top few centimetres of sediment are alive with creatures that can make a snack even of a shark tooth or a calcium carbonate shell. But nevertheless, this is a direct and compelling link between the present and the recent past. Digging or drilling a little further beneath the floor of the sea or a lake reveals yet deeper layers to be found at depths of metres, then tens of metres and then hundreds of metres, each layer older than the one above it, and many still containing fossils. And so on, deeper and deeper physically, and deeper and deeper into the past.

It is one of the great principles of geology that, in a succession of sedimentary strata, the layers beneath are older than layers above. It has been taken to the point at which 'upper' essentially means 'younger' and 'lower' means 'older' when rock sequences are discussed, providing a geometrical track for the arrow of time to fly along. This principle is only applied, it must be emphasized, to the original relations of the superposed sedimentary layers to each other. These layers can be subsequently tilted, crumpled, dislocated, even turned upside down, but their relative original order of rock strata still provides the proxy for time.

This is a principle of great practical application, on the Earth and on other planets. But it is not an immutable law in the way that, say, Newton's laws of motion are. For instance, the sedimentary particles of several successive thin layers of sediment may be mixed up, months and years after they were laid down, by the burrowing activities of animals or the growth of plant roots. What is then buried as a 'stratum' is a diffuse layer made up of inextricably mixed grains that once formed several successive layers, these original layers no longer being separately recognizable. At this level, therefore, time's arrow is blunted. One cannot separate month from month and year from year, but must deal with fuzzy-edged and intergradational parcels of time, each spanning centuries or millennia. But the Earth's history spans millions of years. Therefore, as an approximation, what has become known as the 'Law of Superposition' is an effective and practical means of extracting history from rock strata.

So, with the recognition of rock strata as layers of hardened sediment and of fossils as the remains of long-dead organisms, the road might seem clear to deducing some Earth history. But to understand rocks and fossils it is not so much the land that one must understand. It is the sea.

Turning to the oceans is a logical step in the exploration of this planet. The resting place for most sediment washed from the land is the sea, and most strata represent fossilized sea floors, with their freight of fossilized marine organisms. So, to read Earth history from the stratal record, there is a need to venture into the depths of the abyss. It is not an easy environment to explore. This may not be a simple step, even for technologically advanced interstellar travellers.

Our own civilization has lived by and off the sea for many millennia, and criss-crossed the oceans in frail boats; but only in the last century have humans dived beyond the few metres from which pearl oysters and sponges could be retrieved. Below that the oceans are deep, dark, cold, and possess crushing pressures. To our ancestors, they seemed utterly mysterious and quite unattainable. Did the oceans harbour the kraken-like monsters of legend? Well, we now know that the giant squid comes close, though we still know precious little about the biology of this elusive creature. Does life persist to the bottom of the oceans? That was a closer call, and many early scientists thought not. The abundance of bizarre and phosphorescent life-forms revealed by the searchlights of the early bathyscaphes, therefore, came as no little surprise to oceanographers early in the last century.

The Victorian-era geologists, lacking knowledge of what happens on deep sea floors, were rendered three-quarters blind in their studies. Most rock strata now present on land were lain down in the sea, and often in very deep seas. The study of these strata, lacking contact with a deep sea upon which interpretations could be tested, were led by logic and rules of thumb, both unreliable guides when faced with a natural world that can spring many a counter-intuitive surprise.

One such rule of thumb was the notion of what happened to the masses of sediment pouring into the sea from rivers. Close to the shoreline, that was little problem, for this realm was within reach. Thus, part of the sediment—the heavier and coarser-grained material—stayed close to the river mouth, eventually building up deltas that over time grew out into the sea, much as the Mississippi delta is building out today into the Gulf of Mexico, and the Nile delta—the original delta, recognized as such a phenomenon by Herodotus, over two thousand years ago—is growing out into the Mediterranean Sea. Part of the sediment is redistributed along the shore by waves and tidal currents, to form beaches (on wave-dominated sections of coast) and tidal flats (where tidal forces are concentrated). So far, so good—but deeper?

The rule of thumb—that one might remember from geography lessons at school—was that, once removed from the physical forces that can transport sediment (flowing river water, waves and tides), there would be a natural segregation of sediment, with only the lightest and finest silt and mud particles drifting far from the shore to settle eventually on the deep sea floor.

Faced, though, with real strata, this concept worked only fitfully. Some deposits were found that fitted the paradigm reasonably well: very fine-grained shales and mudstones that showed no evidence of the effects of waves or tides, and that contained marine fossils. These fossils, though, did not seem to include many familiar near-shore or coastal forms: they were mostly strange, delicate organisms that seemed to represent still and quiet conditions. Sedimentary layers such as these could quite reasonably be interpreted as having formed on long-vanished deep sea floors, far from land (though how the sea floors found themselves subsequently hauled up high on to land was another problem). Such ribbons of plausibly deep-water shale now outcrop across the Southern Uplands of Scotland, and the Appalachian mountains. There are fine examples in the Sudetes Mountains of Poland.

But there are huge areas—greater areas—where such strata are finely interleaved with thick beds of coarse sandstone. Much of Wales and southern Scotland, for instance, and large regions within the Alpine and

Appalachian chains are made up of stratal successions like this. These sandstones seemed quite out of place on a deep distant sea floor—how could one transport so much coarse sediment so far? There was evidence that currents had been involved: the sandstones commonly include trains of preserved sand ripples, as might be found migrating on the bed of a flowing river, and sometimes they contained fossil shells of shallow-water type. So perhaps sea level had risen and fallen, dramatically and repeatedly, in these regions? And yet there was no evidence of wave action or tides, forces that only affect shallow-water environments. Things were not adding up, and that was worrying for the whole science, because such strata bulked so large in the make-up of the terrestrial landscape.

An answer to this puzzle emerged quite late in the history of the science, this lateness reflecting the sheer difficulty that humans have had in penetrating below the surface of the water masses that cover so much of the Earth. The answer was born in a kind of improvised bathtub, and helped along by a long memory of a mysterious and destructive event that had happened some decades previously. Thus, off the coast of Newfoundland, telephone cables had been laid on the sea floor to connect North America and Europe. On 18 October 1929, an earthquake struck this Newfoundland coastline. In the hours following the earthquake, the telephone cables went dead—one by one. When they were pulled out of the sea to check what had happened, it was found that each had been snapped by an enormous force that seemed to have travelled rapidly over the sea bed. What could this force have been? There was no ready answer available. It remained as one of many mysteries of the marine realm.

Some years later a Dutch geologist, Philip Kuenen, was experimenting, in the equivalent of a garden trough, what happens when sediment and water mix on a slope. It is an experiment you can repeat yourself by mixing a dilute slurry of water, mud, and sand in a beaker, and (stirring while you do this) pouring it down one of the gently sloping sides of a bathtub. The slurry will transform into a dark billowing cloud travelling along the bottom of the bathtub, accelerating down its flank and then decelerating slowly along its flat floor. This is a turbidity current. Made up of two very different things—solid mineral particles and a liquid—it in effect behaves like one substance, a fluid made more dense and more viscous in proportion to the amount and type of suspended sediment in it.

Eventually (best seen if you have a very long bathtub) this flow decelerates enough for sediment to begin to drop out of it: first the sand and then the mud, to leave a thin graded layer of sediment on the bottom of the bathtub. It is a complex and rather mysterious thing, and fascinating to watch, ever changing its shape, and looking almost like a living creature as it travels. The whole thing is powered by gravity, just like the water in a flowing river, the sediment being kept suspended in the liquid by the turbulent vortices generated in the flow, as leaves are kept suspended by gusts of air in a gale.

A turbidity current is still not completely understood, at least not by human scientists, as it involves that mind-bogglingly complex phenomenon that is turbulence. But it is a wonderfully efficient means of transporting enormous masses of sand and mud a very long way through deep and still water, to accumulate eventually as extensive swathes of sediment, each successive turbidity current leaving a graded layer with sand at the base and mud at the top.

Kuenen and an oceanographer, Bruce Heezen (author, with Marie Tharp, of those iconic—if idealized—National Geographic maps of the ocean floor), subsequently met and recalled the mysterious event off the coast of Newfoundland. They realized that it was a much bigger version of the small-scale phenomenon that Kuenen had produced.

It soon became clear that these events were commonplace; indeed, the norm. Large amounts of soft waterlogged sediment, piled up anywhere at the top of a slope at the edge of an ocean, can be dislodged—by an earthquake, say, or by a passing hurricane—to trigger a turbidity current. Once initiated, these can flow enormous distances, and involve enormous masses: the mobilization of tens of billions to hundreds of billions of tons of sediment in each event is commonplace. Today, the sediment pouring off the Himalayas is carried by the Ganges and Brahmaputra rivers to the edge of the Bay of Bengal. Part of this builds up as the Ganges–Brahmaputra delta (that is, in effect, Bangladesh). Most, though, is carried on in successive turbidity currents across the Indian Ocean floor, cutting deep submarine canyons on the way, to pile up eventually in layers that may still be over a metre thick at two thousand kilometres from the triggering point. Over time, these layers build up an accumulation of sediment known as a turbidite fan. Each event is catastrophic, involving the sudden burial of thousands of square kilometres of ocean floor and everything that lives on it, under a suffocating blanket of suddenly dumped sand and mud.

Yet these are normal catastrophes, typically occurring every few decades or every few centuries on any given turbidite fan. The sea floor is adjusted to them, and so is its assemblage of living organisms: thus, following mass mortality, there is rapid re-colonization. Where turbidity currents spread over the ocean floor, they build up, over geological time, as thick layers of distinctively striped strata, made of alternating layers of graded sand and mud. And where such sea floors are raised above sea level and are transformed into landmasses (such as, say, the Welsh mountains), the characteristic stratal signatures persist, evidence of long-vanished deep seas, to those who can interpret them.

To us, submarinescapes of this kind are effectively alien environments, unlike anything that we encounter in our daily subaerial lives. Interstellar explorers alighting upon the Earth must therefore take time to unravel the particular workings of such phenomena as turbidity currents in terrestrial oceans, and to recognize their long-stilled counterparts within rock strata. They would, though, be able to make comparisons with related phenomena that they would certainly have encountered on other planets. Turbidity currents are part of a broader range of density currents, all driven by gravity and all involving some means of keeping the particles suspended during transport. Avalanches of rock or ice, for instance, which keep travelling for as long as the fragments keep colliding with each other and transferring momentum and kinetic energy from fragment to fragment. Or pyroclastic flows, those most fearsome products of volcanic eruptions, in which incandescent ash particles mixed with hot gas forms the fluid, and a mixture of particle collisions, turbulent vortices, and upward-escaping gas keeps the particles aloft. Or the debris from meteorite impacts, speeding outwards from the crater like a blizzard charged with boulders and molten rock spray.

Each of these produces characteristic deposits. Pyroclastic flow deposits, termed ignimbrites, for instance, rarely show the regular striping seen in turbidites; they typically produce more massive deposits, filling valleys and spreading out over plains, that in places include the traces of giant dunes, whipped along by the intense, short-lived hurricanes of searing ash. Meteorite impact deposits include droplets of flash-melted glass. The telltale patterns of each type of deposit, forensically examined, give clues to the particular processes that produced the deposit, and therefore can help reconstruct the environment in which these processes operated. It is the history game.

Examining modern environments is the best guide to interpreting ancient deposits. The present is a key to the past—but one, though, that does not always provide a perfect fit. Planets change their behaviour as they age. On Mars, there is no modern equivalent of the ancient meandering river channel revealed by that remarkable satellite image. Rivers there ceased flowing billions of years ago. To understand those particular strata one has to go to Earth and study the Mississippi, or the Thames, or the Okavango. Similarly, on Earth, the past includes some landscapes and events with no close modern analogues, as our interstellar traveller-explorers must gradually discover.

But first, particularly as they begin to explore the oceans, they may well be surprised and puzzled by what they find. For there is much in that realm

that is counter-intuitive, and that would have them thinking deeply about how this planet functions as a whole. They will need some ground rules for Earthly planetary behaviour, before the history of that body starts to make sense. It is time that they, and we, encounter the tectonic escalator, and divine the formidable machinery that drives it.

Tectonic Escalator

The crust on this planet behaves most strangely. It must be somehow in motion, for how else to explain the high mountain peaks? Yet, this movement seems not to have annihilated the delicate surface life-forms. There is a subtle engine somehow driving this planet, and finding out what it is should tell us why this planet is so thickly clothed in sedimentary detritus. Perhaps it will tell us also why there is this abundance of living things, and explain much more about this remarkable planet. Our biologists have much to do. Our geologists, for now, perhaps more.

How does it work, this engine that produces the world's strata, those storehouses of an almost infinite history? Our future explorers might be sorely puzzled, for the Earth's motor is quite specific in its mechanism. There is nothing else like it in the solar system, and even reasonably close duplicates of it may be rare among planetary systems generally.

A problem is immediately encountered in any attempt to construct a history of the Earth's life and environments from the stratal archives. For the question will extend beyond simply explaining why the strata that formed on ancient sea floors happen to be present high up on land. Any explorer, in trying to construct a coherent history of the Earth, will find anything but coherence in those rock layers, once they try to put them back into their original order.

For in many regions of the earth the strata are tilted, or are upside down, or are crumpled into huge folds, or have been sliced into segments in which the primary stratal layers are markedly offset from one another.

Some layers show signs of having been recrystallized by heat and pressure, showing that they must have somehow been carried down to great depths below the Earth's surface, and then carried back up again to lie exposed at the surface. The strata of neighbouring Mars, by contrast, have nothing like the richness of the Earth's—but neither do they possess such formidable structural complications. These crazy Earthly stratal geometries, just as much as earthquakes or volcanoes, are indisputable signs of an active planet, in which the seemingly solid and stable crustal surface is, in reality, highly mobile.

Our future explorers should take it for granted that strata are essentially made of sediment that was eroded from topographic highs (say, mountains) and was carried down to topographic lows (say, the floor of a lake or of a deep sea floor). This is straightforward. It has happened, say, on Mars. But on Mars it essentially happened as one cycle, a long time ago, where the highlands represent the eroded areas and the flatlands are an accumulation of the sediment derived from them. On Mars those strata are flat-lying still, after three billion years, after which little of real consequence has happened. But the Earth, phoenix-like, has kept renewing itself, and covering its surface with yet more strata, and so it has faithfully recorded virtually every interval of the last nine-tenths or so of the Earth's four-and-a-half-billion-year history. How so?

Given the basic role of gravity in directing the flow of sediment over any planetary surface, it is clear that the production of strata requires the ongoing production of topography, of areas of high and low ground. The Earth is an exceptionally good topography-maker, with a variety of mechanisms at its disposal.

Sometimes the mechanism is as obvious as the simple filling in of a hole in the ground. Windermere in the English Lake District, for instance, was originally excavated and deepened by the brute force of glacier ice. It is still a lake, but even now it is being filled with layers of mud washed in from the streams that flow into it from the surrounding hillsides. Some time (geologically) soon, it will be entirely filled in, and become dry land. Some fifty metres of strata—perhaps a hundred or so in the deepest parts—may accumulate.

This is piffling compared with the gargantuan thicknesses of rock strata that have accumulated elsewhere. The Grand Canyon shows strata a kilometre thick; it is a spectacular sight, but the thickness is strictly modest—with due respect to Arizonian sensibilities. The ancient hills of Scotland, Wales, the Appalachians, the Urals, and elsewhere commonly show stratal thicknesses well in excess of ten kilometres, and deeper than the deepest sea. More puzzlingly, we can examine some of the rock strata around—to pluck an example at random—the beautiful, and beautifully situated little town of Hallstatt, in the Austrian Alps. The sheer mountainsides around it are uniformly striped, comprising nicely defined stratal layer-cakes, several hundred metres high. Each individual stripe—that is, each stratum, a metre or so thick, though, bears unmistakable signs of having originally accumulated in very shallow water, around an ancient shoreline. What was going on? Was that ancient sea level rising continuously to keep pace with the sediment surface as each layer was laid down? Sea level, globally, can rise—but we do not know of a mechanism that can make it go up by, say, a kilometre.

If the sea has to stay put, more or less, it is the mountain that has to move—almost literally. A little over two hundred million years ago, the place on the Earth that was to become Hallstatt was not in the least mountainous. It was a shallow sea, close to a shoreline. And such it remained, for some ten million years. But over this time, the crust below it sank, by over half a kilometre. Yet the Hallstatt-to-be was not drowned, for just as quickly as the land beneath it subsided, it silted up with limestone-to-be. And the layers of limy sediment, piling higher and higher, compensated pretty well exactly for the sinking of the crust. Hallstatt-to-be kept its head above water, just. It was on a downgoing tectonic escalator. And now that exact spot of the Earth's crust is on its way up again, and around Halstatt-of-the-present, we can see Hallstatt-of-the-past, or rather the thousands of successive Hallstatts-of-the-past, that are exposed on those splendid mountainsides. In another few million years, all those Hallstatts-of-the-past will be gone forever, washed away by rain and rivers. But, as compensation, even older Hallstatts-of-the-past will be exposed to whoever may be around to gaze at them (we contemporary humans may take a sneak preview, simply by drilling a borehole under Hallstatt-of-the-present).

Thus, the crust of the Earth is continuously in motion. The particular sense of motion that is of most relevance to this narrative is up and down. Or, more precisely, first downwards, to bury countless sea floors, and then upwards, to re-expose them as rock layers on some eroding hillside.

Our explorers of the far future will wonder upon the forces that caused the solid crust of the Earth to see-saw to such effect. As they forensically study the evidence, they should eventually realize, just as we humans have lately done, that the Earth's crustal see-sawing is simply a minor by-product of some altogether more dramatic crustal movements, movements so extraordinary that they will seem scarcely credible. There may be real intellectual ferment, one hundred million years from now, for the behaviour of the Earth's crust is one of the marvels of the solar system.

Evidence of indisputable ascent of the Earth's crust is seen in mountain chains: the Alps, the Himalayas, and, perhaps most vividly of all, the Cordillera of the Americas. Leonardo da Vinci realized this, when he found fossil sea-shells on a mountain summit, and put two and two together. *Chains*. The linear nature of the world's great mountain ranges is apparent even from the language used to describe them. For mountains, in reality, do not often form isolated masses, or scattered clumps, the way they do in fairy tales or in Middle Earth fantasies. They stride across the landscape in lines, which is why the early explorers and migrants found them such devilishly difficult things to cross. You could not get around them. You simply had to get *through* them, inching fearfully along vertiginous passes, more often than not. Those intrepid early explorers likely cursed the mountains, were awed by them and puzzled by them in equal measure. Why did the landscape rear up so dramatically? Why were the mountains here at all? Why indeed.

While those first transcontinental travellers struggled, the early geologists puzzled also. For the mountain ranges were not just rocks pushed up towards the sky in thousand-mile-long lines. Those rocks had been intensely crumpled, by massive and unknown forces. And, before that, there was evidence of a strange coincidence: enormous thicknesses of sediment had piled up on the sea floor exactly where the mountains were later to rise up. When the early geologists began tracing bundles of rock strata from the mountain ranges to the more sedate landscapes bordering them, they found that the strata in the mountains were much thicker than those on either side. So, obviously, huge amounts of sediment had been laid down in deep linear troughs in the oceans. That sediment became rock, and for some strange reason those troughs, later, were squeezed mightily to become transformed into mountain ranges. This seemed more than coincidence.

Geologists of a certain age will have been reared upon the notion of geosynclines, the name given to those sediment-amassing troughs. There were different sorts of geosynclines: eugeosynclines, miogeosynclines, orthogeosynclines—the terminology was marvellously elaborate. The geosyncline idea worked well in describing the geometry of the strata that made up the mountain ranges, but it was vague when it came round to explaining quite why all this should have happened. For commonsensical and hard-headed geologists of the time could not seriously entertain the notion that the Earth's crust had moved sideways, by much more than the small amount necessary—some tens of kilometres—to crumple the rock strata making up the mountains.

The idea of much greater crustal movements had been mooted, to be sure, and there was a worrying amount of evidence in favour of it. There was the way that the coastline of South America was eerily parallel to that of Africa. And it was not just the mere outline: the very grain of the continents, the lines of ancient mountain chains, matched up pretty well perfectly. The fossil dinosaurs of those two continents were similar, too. As terrestrial dinosaurs were not, in anyone's interpretation, swimmers strong enough to take thousand-mile sea voyages, the idea of land bridges came into vogue, to make pathways for those energetically migrating saurians. These pathways conveniently later sank, Atlantis-style, into the deep oceans. The necessary pattern of land bridges became increasingly complex to explain the way in which fossils—not just dinosaurs—resembled each other (or did not) in different parts of the world. As more knowledge accumulated of the distribution of fossil animals and plants, a veritable spaghetti junction of former land bridges began to be needed.

Several early-twentieth-century geologists were intrigued by this uncomfortable set of circumstances and wondered whether the continents had, somehow, drifted loose of their bearings and drifted across the globe, at times splitting, and at times joining with other continents. Such a notion would help explain the distribution of the fossils. The most energetic and committed of these was Alfred Wegener, who, in 1912, published a book detailing an impressive array of facts in support of continental drift.

It was generally agreed that Wegener had a point—indeed, quite a few points, and the situation was all very perplexing. But it was still not half as perplexing as trying to explain how continents could plough through the crust of the oceans then to collide to spectacular effect with each other. For Wegener offered no sensible mechanism to shift those continents. And his visionary insights were treated with anything from caution to barely suppressed anger. This seemed to be science fiction, an incredible hypothesis.

Arthur Holmes, a British geologist who had gained a reputation for his early experiments in dating rocks by exploiting the natural radioactivity they contained, and also for his clear scientific writing, did not think Wegener's idea so crazy. He worked out a mechanism that could explain the drift of continents. Perhaps, he surmised, in a paper written for the Geological Society of Glasgow, the partly molten mantle under the Earth's crust was forever in motion, circulating in convective loops, like a cauldron of thick jam being heated from beneath. The Earth's crust might be dragged along with the upper part of the mantle, torn apart in some places, and compressed together in others. Some of his fellow geologists acknowledged this possibility, but, for the most part, still could not be convinced. These were *continents*, after all. Enormously big things. Moving them across thousands of kilometres was surely too far-fetched to be countenanced.

The evidence that suggested that continents had really moved came not so much from any great leaps in theoretical understanding, or from yet more data gleaned from the continents. It was the oceans that did it. For, in Wegener's day, the oceans were vast dark abysses, cloaked by water miles deep, more impenetrable than steel, more mysterious even than outer space. But whatever lay below the deep-sea oozes of the abyssal sea floor, it was pretty well taken for granted that it was just as ancient as the crust of the continents, with a history stretching back millions of years.

Yet there were signs that the oceans were in some fundamental way different from the continents. Early measurements of the gravitational pull exerted by the crust underneath the oceans showed that it was consistently more dense than the rocks that continents were constructed of. Thus, the ocean floors were not just continuations of the land surface that just happened to be deeper-lying; they were made of a different material. And there was their sheer *depth*. For even the crude early depth-soundings taken in Victorian times showed that most of the surface of the Earth lay at two levels: a level around sea level, give or take a couple of hundred metres, and a level at around five kilometres *below* sea level, the depth of most of the ocean floors. There were high mountains, to be sure, but they made up only a few per cent of the land surface. And there were some deep trenches in the oceans, narrow gulfs reaching to over ten kilometres below sea level. Between the two levels of average continent and average ocean floor was, almost everywhere, a relatively steep slope. Strange.

Things became stranger still when specially designed ships began to traverse the oceans, overcoming all manner of technical difficulties to haul drill-cores of sediment and rock from those depths and using sound waves to probe the thickness of the sediment layers covering the ocean floor. It became apparent that something was seriously amiss with the idea that the ocean basins were ancient. For nothing at all ancient could be found. Everything down there was just so . . . young.

The ocean floors everywhere turned out to be geological infants. While some parts of the continents were gnarled ancients, more than three thousand million years old, scarcely anything on the ocean floors could be found that was more than one twentieth of that age. Most of the present ocean floor had, in fact, come into being well after those evolutionary latecomers, the dinosaurs, had passed into oblivion. Drilling, and echo-sounding of the ocean floors, showed that they were made of thin layers of youthful sediment atop dense, iron- and magnesium-rich basaltic rock, dense rock which explained that strange excess of gravitational pull they possessed, compared with the light, silica- and aluminium-dominated continental crust. The sediment layers, also, got thinner and thinner, and more and more youthful when approaching the central parts of some oceans, notably when approaching the ridge that ran down the middle of the Atlantic Ocean, until, at the ridge itself, there is only a light dusting of modern oozes overlying the heavy basaltic rocks. What *was* going on?

It was magnetism that finally led to the key insight. It was known, by then, that the Earth's magnetic field had flipped, in the geological past, repeatedly between the present state, with a magnetic north pole at the North Pole, as it were, and a state where the North Pole became the south magnetic pole. Some rocks, particularly iron-rich volcanic rocks such as basalts, possess iron minerals that align themselves, like tiny compass needles, towards the north magnetic pole of the time when the basalt was still molten magma. Once that molten magma cools and solidifies, all those little compass needles are frozen into place and impart a type of magnetism, termed a remanent magnetism, to the basalt. This remanent magnetism, of countless rock-encased compass needles, is detectable, and measurable by specialized instruments—magnetometers—towed behind ships.

A magnetometer towed behind a ship steaming away from, say, the Mid-Atlantic Ridge, would show that the basalts of the ocean floor at the ridge itself would have magnetite crystals pointing towards the present North Pole. A little farther from the ridge, though, and they would point towards the *South* Pole. A little farther still, they would point towards the North Pole again, and farther yet they would all be aligned towards the South Pole once more. That patient ship would go back to the ridge, sail a few hundred kilometres farther along it, and then try a parallel traverse to see whether it got the same pattern. It did. The Earth's ocean is, in fact, magnetically striped, rather like a barcode. The stripes run parallel to the mid-oceanic ridges, and show ...

The penny had to drop. Harry Hess, an ex-submarine commander, had realized that the ocean crust must be newly made, and Fred Vine, a young PhD student, and his supervisor Drummond Mathews realized that the magnetic data revealed the ocean floor as tape recorder. Their vision had striking echoes of that idea conjured up by Arthur Holmes, almost half a century before. The ocean floor was being continually produced at the mid-ocean ridges, from which it moved apart on both sides, like two mirror-image conveyor belts.

The youth of the ocean basins is undoubted. It is likely that this would be deduced by an extraterrestrial scientific expedition once they seriously began contrasting the overall physical properties of the Earth's continental and oceanic crust: for, technologically advanced, they would surely be probing the Earth's gravitational and magnetic properties, and deploying the likes of radar and sonar. They would thus detect the continent–ocean disparity even before they had really developed the tools (particularly the use of fossils as measures of geological time) to work out the detailed histories encoded within strata. Discovering the magnetic stripes would help them greatly, but the rock property that shows that this—the alignment of tiny magnetite crystals—may well seem (to an extraterrestrial eye) initially so obscure and scientifically trivial as not to merit systematic analysis. Even without this information, the realization that the Earth's oceans have been steadily splitting apart and widening would immediately give rise to deductions concerning whole-planet behaviour. And, of course, there is more than one kind of deduction that can be made.

One option here is that the Earth has been expanding, growing in overall volume and diameter, over the duration of the preserved portions of ocean floor (that we now know to be a little over two hundred million years). This sounds like science fiction. Yet, it is an idea that was briefly in vogue among a small number of geologists in the 1970s, and indeed is still adhered to now, perhaps even more fervently, by a somewhat smaller group. Their vision is a striking one: that the Earth for most of its history was *much* smaller, something like half its present diameter, and did not possess ocean basins. And then, for reasons unknown, it began to puff out like a balloon to reach its present size—and it is expanding still. Indeed, if the alleged current rates were maintained, then the Earth's diameter would be visibly bigger—perhaps 20 per cent greater in diameter—with the continents looking ever more isolated in a sea of oceans (as it were) by the time the interstellar explorers finally arrive.

Crazy? Well, the Expansionists (as the adherents to that theory are known in countries where they have a significant presence) have suggested various lines of evidence in favour of their world-view. That drifted-apart continents can—arguably—be fitted back together more effectively on a smaller-diameter Earth, for instance. And that, if the evidence that new ocean crust has been (geologically) recently created is secure, there remains the problem—insuperable, in their view—of removing old ocean crust, as is necessary if an Earth of constant diameter is to be maintained.

The problem of what to do with old ocean crust would likely puzzle extraterrestrial explorers, especially if such a problem was outside their experience of comparative planetology. The answer to this conundrum is by no means obvious. So, they might well entertain the notion of an expanding Earth, at least initially. But the problems of this idea would soon strike them too; some of these problems would derive from planetary comparisons, even if the comparisons are made only within the confines of this solar system.

For the Earth, in this interpretation, would be the only expanding body in its system, at least as regards the rocky planets and their satellites (when it comes to the gaseous planets, such as Jupiter and Saturn, it would be difficult to prove or disprove expansion, for these do not possess a fixed surface frame of reference). The Moon, for instance, and Mars have solid surfaces that are ancient and easily visible and mappable, and have gone through the period of time in which the Earth is supposed to have swelled up without themselves showing so much as a hint of expansion. And what would the expansionary mechanism be? A general reduction in the strength of gravitational fields, allowing tightly compressed mineral structures in the Earth's interior to expand, and thus to occupy more volume? Again, this should affect all the planets (and solar systems within galaxies), not just one, while a general reduction in gravity would also play merry hell with Newtonian orbital mechanics. And there is also the matter of the effect upon the Earth's climate and life support systems of a relatively late, sudden appearance of oceans, in the last five per cent or so of terrestrial history. As this terrestrial history is slowly pieced together, the sustainability of the Earth's ecosystems over such a revolution would inevitably loom as an ever-greater problem.

Proponents of Earth expansion might consider that such problems are outside their terms of reference, and that they are only concerned with the evidence they see, it being up to others to come up with an all-embracing theory to explain all of this. Well, our future archivists may examine this problem from a broader perspective, and so might only—after some consideration of the matter—retain the hypothesis of a single, markedly expanding planet within a solar system as a last resort, if all other hypotheses failed. So, they will be looking to see how a planet might dispose of substantial amounts of old and unwanted ocean crust, quietly, discreetly, and without terminally disturbing the delicate and all-too-vulnerable network of living inhabitants of that planet.

Destroying sections of crust is no easy business. Especially as 'crust' here turns out to be not just the thing in its technical sense, this being the Earth's surface rock down to the Moho (as geologists call the Mohorovicic Discontinuity, below which the Earth's mantle lies), which is some 10 kilometres thick in the oceans and typically 30–50 kilometres thick in the continents. It means something termed the lithosphere, which is up to 50 kilometres thick below oceans (and some couple of hundred kilometres thick below continents). The lithosphere lies above a layer—the asthenosphere—that is significantly physically weaker because it is slightly closer to its overall melting point. It is a large amount of rock to dispose of, and this cannot be done without leaving some tell-tale clues at the Earth's surface.

Clues there are: linear zones of extreme topography and intermittent mayhem. The most obvious of these today is the Pacific 'Ring of Fire', bordered to the east by the American Rocky and Andes chains, and to the west by a more complicated network that snakes through by Japan, Indonesia, Sumatra. Here, there are earthquakes, and trains of the more violent volcanoes. The rocks, too, are notably crumpled and sliced, and some show signs of having been plunged to great depths and then equally rapidly brought back to the surface again.

These are zones where the ocean floor is destroyed, entire lithospheric segments descending into the depths of the Earth. Their downward journey starts at the deep oceanic trenches, and their difficult, friction-ridden passage downwards is marked by powerful earthquakes. The continents do not have to plough through the oceanic floor. Rather they are carried atop the continually moving lithosphere. They forever drift, like so many gigantic, stony Flying Dutchmen, as the ocean basins slowly, inexorably open and close.

The Earth's crust, thus, in this radical model, is far from whole. Instead, it is crazed over with cracks like Humpty Dumpty's shell, the cracks marking the edges of the slowly drifting tectonic plates. The crust underlying the ocean floors simply represents the sweated-out expression of the giant convection cells which roil slowly through the hot, partly molten stuff of the Earth's interior, several hundred kilometres down. The continents are the scum which, once formed, are always too light to be dragged back into the Earth's interior, and so are fated to be truly immortal. Everything fits together neatly in this vision of the world as a huge, slowly simmering sphere with a mobile outer surface.

And, continental drift is now a fact. The Atlantic Ocean, that gap between Europe and North America is ever widening, growing at the speed of a human fingernail. This has been established, proved beyond doubt, by the almost impossibly precise measuring devices carried by satellites in orbit. The means by which this happens, the hypothesis of plate tectonics, is, though, still just a hypothesis. But it is a hugely successful hypothesis, one that accommodates pretty well every geological observation currently being made, be it to do with volcanoes, earthquakes, or the distribution of today's animal and plant communities.

Given the scale of the crustal rearrangement, it is also a remarkably gentle process. True, it kills people every year, by earthquake, volcano, and tsunami. But this needs placing into context. The Mid-Atlantic Ridge, for instance, marks a line where the Earth is literally splitting apart. This planetary-scale, ever-opening scar is marked, at geological timescales, by continuous outpourings of magma. On human timescales, though, Icelandic people live astride that ridge. They farm and fish and generally live prosperously (if, as a visitor, you wish to see how prosperously, just try buying a beer in Reykjavik).

The collision zones of plate tectonics provide more serious violence. Today, a thousand kilometre-long, 50 kilometre-thick section of Pacific Ocean floor is being forced down beneath South America. Entire ocean islands here are simply sliced into pieces, scrunched, and mangled. This is the territory of big earthquakes, and of the more lethally explosive volcanoes. Ideally, one should not live near them. In an overcrowded world, though, many people do not get the choice, and the volcanic soils are fertile, so today's full belly outvotes tomorrow's disaster.

Plate tectonics also provides terrestrial creatures such as ourselves with a dry home. The continents, the accumulated scrapings of the whole messy process, float eternally high on the backs of the ever-moving plates. If plate tectonics had not developed on the Earth, then there would be no continents. Could humans, or human-like creatures still have evolved? As web-footed merpeople, perhaps, gliding elegantly through a deep calm global ocean, rather than in the current model as clumsy landlubber humans. But this is doubtful. Without the land, there would be no rivers, and without the rivers, no large-scale cycling of nutrients into and through the oceans. With such a scenario, it would be hard to imagine gill-bearing humanoids appearing. Evolution might, perhaps, have got as far as some tough bacteria fighting over a few nutrient scraps.

There is another reason to be grateful for the shifting plates. The Earth generates a great deal of internal heat, because of the radioactivity inside it. That heat has to escape somehow. Conduction is not fast enough, because the Earth is simply too big. So the almost-but-not-quite solid interior of the Earth is slowly convecting, and this motion drives the overlying plates, which in turn provides a regular means by which magma can reach the surface to release its heat.

Imagine an Earth without such a smoothly running heat exchange device. In fact, we do not have to imagine. Venus is a rocky planet, and about Earth-sized. It does not have plate tectonics. It seems to have evolved a different, and much less agreeable, form of heat release. From the amount of the meteorite impact craters on it, its land surface seems to be about half a billion years old throughout. That is an unusual pattern, and requires explanation.

Half a billion years ago something seems to have happened. That something has been called 'resurfacing', when Venus created for itself a brand new landscape. The scenario involves half a billion years plus of pent-up heat suddenly coming to the surface in a planet-wide maelstrom of lava. The whole planet, in this interpretation, literally turned itself inside out. This would have been bad news for any recently evolved Venusian lifeform, no matter how superbly it might have been adapted to the immediately pre-existing Venusian landscape.

What makes the difference between these cosmic bodies, between planetary Jekyll and planetary Hyde? A good deal of the answer might be plain, familiar water, which the Earth has in abundance. It is a highly effective lubricant, that can help even 100 kilometre-thick slices of crust slide past each other. The water on Venus boiled away a long time ago, and those thick swirling clouds are not steam but hot vitriol. On a dry planet, plate tectonics, seemingly, cannot work: there is just too much friction. And so steady heat release seems to have been replaced by intermittent Armageddon.

One can make a good case that we really are living on the best of all possible worlds. The good Doctor Pangloss was right all along. It is amazing how it all works: the Earth really is a dream machine: unique among its neighbours and perhaps a rarity at the galactic level.

So why, given all the evidence that could be mustered for this model of an Earth machine, did the idea of drifting continents take so long to catch hold? For, in hindsight, the evidence, painstakingly assembled by Wegener and by his few successors such as Alexander Du Toit, could not be explained in any other way. The very outrageousness of the driving mechanism, plate tectonics, by which continents move across the Earth's surface is one factor. Like the bumblebee, this idea should not fly, yet it does. Yet another factor must have been the passion with which Wegener and Du Toit argued their case. They were converts. Most scientists, though, are a sceptical lot, and dislike conversion to any idea, especially a new and unproven one. Anything other than cool and dispassionate arguments, with due weight given to possible counter-arguments, is a priori suspect. One critic termed Wegener's stance 'auto-intoxication'. And so most scientists of Wegener's day picked upon any flaws in his arguments.

Flaws were not hard to find. For instance, Wegener originally cited as evidence of continental drift the fact that Ice Age deposits—boulder clays, sands and gravels and the like—reached about as far south in Europe as they did in North America, and hence suggested that these land masses had broken away from each other only a few thousand years ago. This was nonsense, and Wegener should have known it. The ice sheets marched so far south in both continents simply because the climate that controlled them operates, by and large, in latitudinal belts across the globe. Wegener's good arguments—which were most of them—were rendered suspect by the few that were bad.

Still, Wegener was, in essence, correct, though in this he had good luck as well as good judgement. The continental drift that he posited spawned the hypothesis of plate tectonics, and the mechanisms of plate tectonics control most of the Earth's geology, including the production and accumulation of strata.

So how, in this new context of an active and functional Earth, do strata accumulate to leave their message into the far future? Overall, the tectonic plates of the Earth's crust can either move apart, continually creating new ocean floor between them, or they can collide, in which case one plate has to slide beneath the other; or they can simply move past each other, much as the two continental plates either side of the San Andreas Fault in California are moving past each other. This is the framework that defines the Earth's sediment traps, and determines the production of the strata in which earthly history is written.

The really good sediment traps are not, somewhat counter-intuitively, produced *very* close to a tectonic plate boundary. Take for example the Mid-Atlantic Ridge, where the Atlantic Ocean is being pulled apart. Along this line, Iceland is the most visible manifestation of this, for most of the Ridge lies a kilometre or so below the sea. Iceland is currently pushed so high because it lies directly above a fountain of partly molten rock termed a mantle plume, ascending from hundreds of kilometres deep in the Earth, which is slowly forcing its way upwards and distorting the crust immediately above it. Still, any piece of Iceland is being carried sideways, either to the east or the west, depending on which side of the central fissure it happens to be. Eventually it will slide underwater and be carried into the deep ocean. So far, so good. The rocks that represent the two halves of the former island will become ever more widely separated, and then be covered by deep ocean muds, over the next few tens of millions of years.

These deep ocean muds can tell some marvellous stories in their own right. But geologically they are strictly short-term. They do not travel upon this ocean floor conveyor belt forever. Nor, at the end of their journey, do they tend to be raised to the surface by simple and gentle tectonic uplift, in a manner that might preserve the coherence of the intricate stories preserved within them. Eventually, these ocean floors and their stratal cap come up against an immovable object, such as a continent that, because of its greater buoyancy, must perforce stand its ground.

Oceanic crust is forced downwards in such regions, with deep oceanic trenches marking the lines along which the ocean floor slides beneath the irrepressibly and everlastingly buoyant crust of the continents, and is forced hundreds of kilometres down into the Earth, where any original structures are simply crushed, sheared, or melted out of recognition. The material on such a downgoing slab is almost certainly on a one-way ticket to cosmic oblivion. And here almost all of those strata-bound stories are lost, shredded as effectively as the throwing of books into a furnace. Some, perhaps, might be scraped up, like wood shavings curling out of a carpenter's plane, to be plastered against the side of a growing mountain chain. These fragments are all that survive of the deep ocean records.

It is the continents that provide most of the really long-term stratal records, that reach far back into the Earth's history. In the immediate collision zones, enormous thicknesses of strata may accumulate—those 'geosynclinal' ocean trench sedimentary deposits. These, though, are comprehensively mangled as mountain chains rise up, driven by the pressures generated where tectonic plates converge. Less obvious, but more widespread, are the wrinkles which extend from the immediate zones of plate impact and around the stretched area of crust where plates are pulling apart. In these, sections of crust are either going up wholesale, creating uplands, or sinking down, creating great crustal depressions in response to the stresses and strains caused by the movement of tectonic plates.

Now, a sinking piece of crust is nothing more nor less than a hole in the ground. And holes in the ground, in nature, seldom stay unfilled. They fill with water and silt up, become filled with sediment eroded from any adjacent part of the Earth's crust that is moving upwards at the time. This generalized see-sawing of the crust produces, in effect, tectonic escalators, taking tens or hundreds of million years to make kilometre-scale ascents and descents.

Britain, for instance, has been nudged by the opening of the Atlantic Ocean on one side, and the building of the high Alps on the other. A hundred million years ago and more, slow earth movements began carrying silted-up continental shelves and vast deltas, in which dinosaurs were entombed, a kilometre or so downwards. And then they came back up again. Dinosaur fragments occasionally surface in these exhumed shelves and deltas (part is now called the Weald of southern Britain) and may be chanced upon by the lucky prospector.

These tectonic escalators are remarkably common. It is hard, in fact, to find examples of the Earth's crust which are not moving upwards or downwards to some extent. The land tends, by and large, to be on an upward escalator. That is why it is land, of course, and above the sea. But the 'by and large' conceals quite a few exceptions (exceptions which will guarantee, incidentally, our species' geological immortality). For a piece of land can stay above sea level not only by moving upwards en masse, but also by being silted up just as rapidly as its crustal foundations are sinking. Examples of this fine balance include most of the great delta plains of the world: the Mississippi, the Nile, the Ganges–Brahmaputra, the Rhine, the Yellow and Red rivers. All, today, are thickly populated by a humanity exploiting their fertile soils and abundant water resources.

In the areas which have been steadily subsiding, and where sediment has been piling up, layer by layer, over tens of millions of years, there is also a positive feedback at work. The more sediment piles in, the more the crust is further pressed down, simply because of the sheer weight of deltas and silting-up coastal plains. A descending tectonic escalator is thus given a powerful additional push. The stage is set, then, to produce a continuous record of the strata that accumulate in these regions. These are pickling jars for the animals and plants that live on the surfaces of those deltas. And for the cities, too, that have been built on them.

Imagine a piece of crust that is being pushed slowly downwards by tectonic forces. It is next door to a piece of crust—say a mountain range—that is being pushed upwards. That is a common enough situation, and one that can be modelled in the comfort of your living room. Next time you sit on a beanbag, try pushing down on one part of it. It will give way quite satisfactorily, but the displaced beans will cause the adjacent parts of the beanbag to bulge upwards. The Earth's crust acts a little like that beanbag, and the same principle applies—if one part goes down, other parts must go up to compensate and vice versa.

There is a clever twist to this compensatory uplift or downwarp, which applies to the Earth's crust, though not to beanbags. For the crust that is going up is eroded by wind and rain, and broken down into grains of sediment, which are carried by rivers into the hole where the adjacent crust is descending. This sediment, once it is piled up to thicknesses of a few tens or hundreds of metres, is heavy (remember when you last tried to carry a bucketful of sand?) and presses down on the crust that it lies on. That crust, being pliable on length scales of hundreds of kilometres and timescales of millennia, simply gives way beneath the weight and sinks downwards even further.

This giving-way of the crust has two effects—it deepens the crustal depression into which the river-borne sediment is pouring, and thus amplifies its action as a trap for incoming sediment. And, by means of the squeezed-beanbag mechanism, this gives an extra upward push to the parts of crust that are going up—which in turn give the wind and the rain more of a target to aim at, and so these generate more eroded sediment—which is then carried by rivers to the downgoing part of the crust, and depress it still further. Meanwhile the uprising mountain range, made lighter by the removal of all that heavy sediment, rises upwards even farther. . . . And so on.

It is a simple mechanism, but effective. The transfer of millions of tons of sediment is both a response to patterns of crustal upwarp and downwarp, and a means of amplifying those patterns. This process is called positive feedback, and it is the Earth's way of making a mountain out of a molehill—and, to ensure symmetry, a chasm out of a molehole. And, incidentally, it is a way of generating thick successions of sedimentary strata within the crustal depressions.

Let us find some examples. The Mississippi delta is a good one, currently being the structure upon which New Orleans is built. The Mississippi delta is not only where the Mississippi River empties into the Bay of Mexico. It is where all the sediment carried by the Mississippi River from the interior of North America is being dumped at the edge of the Bay of Mexico. That sediment is arriving thick and fast, and its enormous mass is pushing down the crust upon which it is resting. Large parts of the rest of North America are slowly rising in compensation.

The rates of delta growth and subsidence are impressive. A lobe of the Mississippi delta can grow outwards—create new land—at several kilometres a century, while layers of mud and sand from tens to hundreds of metres thick have piled up in the last few thousand years. Space to allow that thickness of strata to accumulate has been created by subsidence of the land surface. Not all of the subsidence relates to the tectonic see-saw. Part of it is due to the weight of the surface sediment squeezing water out of the more deeply buried muds, causing the mud particles to be squashed together more tightly and the mud layers themselves to get thinner. Thus, a layer of soft, fluffy mud at the surface might be made up of just one-quarter mineral and three-quarters water; once it has been buried beneath a few tens and then hundreds of metres of newer sediment, most of its water content may be squeezed out, and so the layer will compact down to half or less of its original thickness . There has, more recently, been a human factor in subsidence, too. The last few decades has seen a good deal of water, oil, and gas extracted from beneath the Mississippi delta. This has created underground voids that are filled by further subsidence of the surface.

Add that to the wholesale depression of the crust beneath the delta, stemming from the weight of the sediment itself. The net result is that the coastal regions of the Mississippi delta—Louisiana, essentially—are steadily, inexorably, and unstoppably sinking, at about a centimetre every year. The sinking will be balanced by fresh sediment arriving to keep this area built up above sea level. If the rate of sediment influx is greater than the rate of sinking, then the delta will build out into the sea; if it is smaller, then the coastline will retreat, and the sea will advance to cover what was land (not quite dry land; mostly swamp in the case of the Mississippi delta). But any individual level within this will be buried deeper and deeper.

This delta is not alone as an example of downgoing, silting-up crust which is managing to keep its ground surface above the sea—just. There is the Red River delta of Vietnam, the Nile delta, the Makaram, the Po delta of Italy, the Ganges, and many others. These are all major, currently active sediment traps, all amplified by the positive feedback of the beanbag effect.

Examples from the geological past, too, are legion. These are, quite simply, successions of strata that represent ancient floodplains, deltas, and coastal plains. In Britain there is a fine selection: the time-honoured Old Red Sandstone, a 400-million-year-old river system; 50 million years later, the swamps of the Coal Measures; another 50 million years after that, the arid playas and deserts of the New Red Sandstone; 100 million years after that, the broad delta of the dinosaur-haunted Wealden rocks, bearing countless bones and trackways of these reptiles. In the US, there is the Catskill delta, more or less a time-equivalent of the Old Red Sandstone, though one that formed on the other side of the proto-Appalachian mountains.

Each of these represents strata that can be over a kilometre thick—some up to *ten* kilometres. And these strata are nothing more than a virtually limitless succession of buried landscapes, stacked one on top of another. Each of these, upon being buried, preserved evidence of the life that inhabited it.

On the other hand, regions of upward escalator are extensive today, and places here are destined for oblivion, as the crust on which they are built is pushed upwards ever higher towards certain destruction through erosion. We know, pretty well, how today's tectonic escalators are behaving. They include most of the Earth's mountain ranges—though these can, here and there, include local sediment traps, where parts of crust subside, often bounded by fractures in the crust. There are broader regions, too, such as north-west Britain, rising as south-east Britain and the

North Sea (the latter a particularly fine sediment trap, active these last hundred million years) slowly sink.

This is the machine that has functioned without pause since plate tectonics began on Earth, to create a stratal record that extends back, almost continuously, for four billion years. From it we know how oceans, atmospheres, landscapes and life have evolved over that time (the first half billion years of Earth's history, without such a record, is largely a blank). This long record has rules, or at least deeply ingrained patterns, that can allow explorers—ourselves or our future excavators—to predict what can be preserved and where.

There is another mechanism at work, though, that acts with tectonics to control the stratal record, to determine which parts of history are preserved, and which parts are effaced even as they occur. This is a more subtle mechanism, and the interplanetary explorers that we posit as the Boswell to our Johnsonian extravagance will be hard pushed to discern its action, particularly in their early, pioneering days, even though at least part of the mechanism has some magnificent extraterrestrial counterparts. Later, as their grasp on Earth history becomes ever firmer, this mechanism will become ever clearer. So let us consider the vagaries of Neptune's domain a little. It is the wild card that may decide the ultimate fate of many a part of our empire.

High Water, Low Water

The three phases continue to be a puzzle. Liquid water, water vapour, and, on some of the higher peaks, a little water ice. How long has this been so? It seems the optimal arrangement for the peculiar biology, yet the central sun here has, of course, evolved. So how has such stability been maintained? One might wish for someone to have put in place a time-series water-phase recorder in place at the birth of this planet. It might explain much, if we but had one.

In almost everybody's natural lifetime, the sea is one of the great unchanging certainties of life. There is land; there is sea; and in between is that magical place, the seaside, which is sometimes knocked about a bit by the waves, but always manages to recover for that next idyllic summer.

There are, one remembers, those faintly disquieting legends, about a remarkably well-organized and ecologically aware person called Noah, and about a Deluge. But these, of course, should not be taken seriously. They were a jumpy and superstitious lot, our ancestors, always prone to making up scary stories. It was a good way to keep the children in order.

With a longer perspective, things seem a little different. Take any one location on the globe, for instance. Track it over millions of years. At that one location, there may be a change from deep ocean, to shallow sea, to a shoreline, and thence to terrestrial swamps and flood plains. And then, perhaps, to the absence of evidence, a horizon of absolutely no thickness at all within a succession of rock strata, in which a million years or a hundred million years—or more—may be missing, entirely unrecorded. It is that phenomenon called an *unconformity*, all that is left of the history of a terrestrial landscape pushed up into the erosional realm. On that eroding landscape, there may have been episodes of battle, murder, and sudden death among armoured saurians, of fire, flood, and storm, and of the humdrum day-to-day life of the vast vegetarian dinosaurs, chewing through their daily hundredweights of plants. Of this, no trace can persist. Only when that landscape is plunged again towards sea level, and begins to be silted up, can a tangible geological record resume.

The Earth's crust, as we have seen, is malleable, can be pushed downwards or thrust upwards by the forces that drive the continents across the face of the globe. Many of the sea level changes that can be read in the strata of the archives are of this sort, and mark purely local ups and downs of individual sections of crust, with no evidence that global sea level was anything other than constant.

Homo sapiens extends back some 160 000 years, a short time span for a species. Even so, the early part of this history includes encounters with changing sea level—hence, perhaps, the Noah legend, filtered through stories repeated and embellished, generation after generation. More recently, as human civilization developed a written history, over the last five thousand years or so, global sea level seems not to have changed by more than a few tens of centimetres. This has allowed the growth of the great seafaring cities: Alexandria, Carthage, Venice, Constantinople, Amsterdam, Genoa. Such sites were arguably among the most effective springboards for humanity's takeover of the Earth. These cities flourish still, as part of our familiar land–sea geography, even if they—and this geography—are now on borrowed time.

Our far-future explorers will arrive on an Earth with a land–sea geography different from today's, as plate tectonics will have shifted the continental masses to new positions. But that future Earth will have well-established, seemingly stable coastlines—wherever those coastlines might happen to be. The direct observations of such visitors to Earth will be, at least initially, too short-term to deduce whether sea level on this ocean-dominated planet changed at all, let alone whether it showed predictable trends or patterns. The land–sea arrangement, and the shorelines, will appear to be a constant. The slate here will start blank.

But, as planetary explorers, they will not expect constancy. Planets evolve through time, and they will have come across many worlds akin to Mars, with its dried-up shorelines, and Europa with its fractured ice surface overlying a globe-encircling ocean. The delicately balanced oceans of the Earth—covering most of the globe but not all of it—will, from that perspective, seem a phenomenon central to the maintenance of an environment where complex life-forms can appear and flourish. The controls on those water masses, and their history, will form a prime research target, before these explorers even get any inkling that a long-vanished intelligent civilization had once colonized this Earth. Later, they will realize just how crucial the behaviour of those oceans was to preserving the remains of a terrestrial civilization.

Where will they begin to see evidence that the oceans had changed their level? They might, by closely looking at the future Earth's geomorphology, be able to discern that some changes in sea level had recently occurred. Cliff lines, say, may have been recently drowned or uplifted. But it would be hard to say whether such changes were due to the land level changing because of tectonic forces, or because of the sea changing its level. And as to what might drive sea level changes . . . well, that might be even more obscure, especially if, as is likely, the Earth would have returned to its long-term climate state, the global warmth it enjoyed when the dinosaurs roamed the Earth.

To pursue this kind of narrative, our explorers will need to have gained some familiarity with the peculiar tectonics of this planet, and have developed some facility at reading the information encoded within terrestrial strata. They will, in short, need to retrace the steps taken by human geologists, and decipher the stratal archives. They will find that the story of this simple parameter—how high or how low sea level is at any time—will touch upon virtually every other surface process on this planet.

From virtually the earliest days of organized geology, a century ago and more, it was observed that some of the trends observed in ancient sea level changes seemed to be repeated all around the world simultaneously, as far as could be judged. An example is the distinctive Chalk deposits that today form the white cliffs of Dover—and also occur widely across the world; these mark a time when large parts of the continents were submerged under water. Such examples could, of course, have been coincidences, in which large sections of widely dispersed crust happened to go up and down simultaneously. But it soon became obvious that a far more likely explanation was that global sea level itself could rise and flood across the low-lying parts of continents, or it could fall, leaving a sea floor high and dry.

How so? For the amount of water in the oceans seems, on the face of it, to be a long-term constant. There are ways, though, of changing sea level simultaneously around the world. Some of these mechanisms can even keep the amount of water in the oceans constant, which is apparently even more of a riddle.

One answer to this riddle is tightly linked with plate tectonics, involving the vagaries of the slow currents of viscous rock, mostly solid but part-molten, deep inside the Earth. The pattern and strength of these deep Earth currents control the formation of oceanic crust at mid-ocean ridges, as molten rock wells out of uprising deep rock currents and solidifies into the basalts that make up the ocean floor. These mid-oceanic areas are ridges not so much because the upwelling magma/rock is pushing them up; rather, they are ridges because the rock is hot and is therefore less dense, and so rises higher. As this rock is carried further from the mid-ocean ridge by the actions of plate tectonics, it cools and becomes more dense—and so the whole ocean floor sinks ever lower.

Now—and here is the clever part of the mechanism—increase the rate of production of ocean floor at the mid-ocean ridges, and the ocean floors on each side of the splitting ridge will drift apart more quickly. The still-hot, expanded rock will take up a greater area of the ocean. The ocean floor will become that bit shallower, on the whole, and so the seawater above that sea floor will have no place to go but up. Sea level will rise, and so what had been low-lying terrestrial landscapes all around the world will be flooded by the sea.

At other times, the rate of ocean floor production, overall, around the world, has slowed down. The rocks on either side of a mid-ocean ridge moved apart more slowly. Thermal contraction took place that much closer to the mid-ocean ridges, and the water settled back into the more capacious ocean basins. Sea level thus fell, and what had been shallow sea became land.

It is a neat trick, which explains some of the overall similarities shared by the world's stratal successions. The mechanism, though, is a little slow. The reorientations of the world's ocean ridge systems, and hence the rises and falls of sea level, normally take millions—or tens of millions—of years to accomplish.

A different mechanism involves considering adding water to, or subtracting it from, the oceans. Water can be taken out of the sea, via extra rainfall—but where does it go? Some can be stored on land, to fill lakes and swamps a little fuller. Our species is helping this process along quite effectively, by damming a substantial proportion of the world's rivers, and creating lakes—that is, reservoirs—where none had existed previously.

Reservoirs, though, compared with the vast mass in the oceans, are but a fleabite. The effect can probably be counted in centimetres of sea level, and in any event is probably offset, today, by humanity's insatiable thirst for water, pumped out of groundwater reservoirs pretty well everywhere. This water comes out of the ground, makes a brief stopover in irrigation ditches, factories, and domestic plumbing systems, and then is dumped into the rivers and thence to the sea.

Then there is the simple expansion of water as it warms. As an old-fashioned thermometer shows, this can be an effective mechanism. In the warmer world of the Jurassic and Cretaceous periods, some twenty metres of the higher sea level of those days may be ascribed to thermal expansion. This mechanism takes a little time, as any extra heat at the surface takes thousands of years to spread through to the depths of the oceans, but it is predictable, indeed inevitable, if atmospheric warming takes place.

There is, though, a yet quicker and simpler mechanism to hand, and one with a long track record of changing sea level with a speed, and by an amount, that can be entirely disconcerting. There is a lot of water stored on our land masses, handily above sea level, in convenient frozen form, and all too easily convertible to liquid. The colossal ice sheets that currently squat over Antarctica and Greenland are kilometres thick. These have grown by taking water out of the sea and piling it up, in solid form, on the land. They currently contain enough water to raise sea level by around seventy metres, if completely melted. There is nothing intrinsically outrageous in considering whether all of the world's major ice caps might one day melt. Polar ice caps were, after all, virtually non-existent in the heyday of the dinosaurs—and, incidentally, for much of the rest of geological time. There is a good chance they will have shrunk to little or nothing one hundred million years from now.

So the question can be turned on its head. Why so much ice? Why now? Here we tread on speculative ground, faced with the monstrously difficult question of global climate, and how the Earth's surface has got hotter and colder at different times. Here, more respectable accounts, coughing discreetly, might draw a veil. But let us plough on, and approach the tender shoots of climate theory with the hobnail boots of blunt enquiry.

To start with, one remarkable, astonishing fact: over a span of at least four billion years, the Earth's surface has been just very nicely, comfortably warm, warm as toast, warm as toes by the fireside. The temperature range of the surface of the Earth has remained astonishingly narrow. The oceans have never boiled away, nor have they ever frozen over entirely (with one possible exception, some 700 million years ago, when the Earth, briefly, may have—there is serious debate here—become a giant snowball).

The mystery of how the Earth has remained within the comfort zone deepens when we consider that neither its external heat supply nor its internal thermostat have behaved with any constancy. There is the puzzle, first, of the faint early sun. Four billion years ago, our sun would have provided (if its workings are understood properly by astrophysicists) only some 75 per cent of the heat and light it emits now. By 2.5 billion years ago this would have risen to a little over 80 per cent, from when it has risen, very gradually, to its present level. Through all this, the Earth's surface temperature has stayed steady. That is . . . unexpected. Somehow playing a role here is the fickle behaviour of the Earth's atmosphere, the windowglass of the global greenhouse, that controls how much of the sun's heat is retained at the surface and how much flies back out to space. The atmosphere has changed its composition quite radically over the eons. Three billion years ago, it had virtually no oxygen, but contained high levels of carbon dioxide and methane. These are greenhouse gases, trapping infra-red radiation radiated by the Earth. Their higher concentrations then provided a warmer blanket than we have now.

It might be, of course, that the changes in the sun's output and the atmospheric thermostat have always neatly cancelled each other out, either by sheer chance or, as the scientist James Lovelock persuasively argues, with the active collaboration of the totality of living organisms on earth, acting as a giant environmental regulator, a superorganism he calls Gaia. This entity cannot do anything about the sun's output, but may somehow manage to modify (say) atmospheric composition and cloud cover to trap just as much of the sun's heat as necessary to keep conditions just comfortable. Just comfortable, that is, for that superorganism's wellbeing. Gaia, in this interpretation, looks after herself.

This is heady and controversial stuff, but let us just be thankful, for now, that the Earth's history has run so smoothly. One has only to look at our planetary neighbours. Venus, besides being dry enough to stifle plate tectonics, is a nightmare of a runaway greenhouse, implacably hostile to any form of life that we can imagine. The atmosphere is crushingly thick, mostly of carbon dioxide, devoid of oxygen or water but laced with vitriol, a blanket that keeps the planet's surface hot enough for pools of molten lead to form. There is no evocation of sensual earthly delights there. Mars, meanwhile, was probably mild and hospitable once, a few billion years ago, with liquid water running over its surface. Now, its atmosphere has pretty well drifted out into space, and its surface is colder and drier than the Earth's prime refrigerators, the dry valleys of Antarctica, those chilly hells where seal corpses have lain, freeze-dried, for millennia.

So the Earth has somehow stayed marvellously well regulated, virtually throughout its existence. But, every now and again, its temperature controls have changed, and ice has built up over the poles, and begun to push into mid-latitudes. One can take the last half-billion years, and studiously avoid what was almost certainly the greatest glaciation of all, christened the Snowball Earth, some 700 million years back. To plunge into this controversial episode would take so many pages that the thread of our story would be quite lost.

For the first 150 million years, then, of our chosen half-billion-year span, the Earth was essentially a hothouse. There was, true, a brief icy interlude at the exact boundary of the Ordovician and Silurian periods of geological time, 440 million years ago, as far as we can estimate. The main icy pulse lasted only a million years or so, with some phases of chilliness fore and aft. These left splendid boulder clays and ice-scratched surfaces in, for example, North Africa (then positioned at the South Pole). The main pulse also devastated the communities of marine animals, with many species not surviving the refrigeration.

Another 100 million years of global warmth followed, while animals and plants left the sea and colonized the land. The colonization, by plants in particular, was so successful that the land greened, almost explosively. Mighty forests grew on—and here we have a geological accident—perhaps the biggest coastal swamp the world has ever seen. Stretching from North America, across Europe to Russia and then China, it nourished, and then buried, generation after generation of strange and primitive swamp trees, which were buried and converted into underground layers of coal. That coal, of course, simply represents the wholesale burial of carbon, extracted from the atmosphere by the photosynthetically active swamp plants. Levels of carbon dioxide in the atmosphere plummeted during that orgy of plant growth. The Earth's temperature dropped in a phase that lasted 50 million years or so, as the polar ice caps grew huge once more.

The chill had abated in the dinosaur-thronged Mesozoic Era. While *Tyrannosaurus rex* prowled around, terrorizing the inoffensive *Triceratops*, the Earth, land and sea both, was more or less evenly warm from north to south. If there were any ice caps at all—and this point is much debated—they were minuscule.

Then the Earth chilled again, into our very own Ice Age. The slide towards our Ice Age was more like a roller-coaster. The evidence as to *what* happened, broadly, is reasonably secure. As the *why*; well, that is another matter altogether. But on to the facts. The last dinosaurs lived in global warmth. Then came the troubles that removed them, and much else, from the Earth. A chill set in. Not a glaciation as such, but the fossil shells, say, a few million years after the great extinction, resemble, here and there (at high latitudes particularly), the coolwater forms of today.

Then, a few more million years later, temperatures soared to levels higher, if anything, than those in the Cretaceous (this was an episode of some contemporary resonance, and we shall return to it). Even when that abated, London was surrounded by a subtropical landscape, thick with palms, through which crocodiles stalked. And then, quite abruptly, some thirty million years ago, another chill. This time it was a plunge, particularly, in the temperature of the deep ocean waters. These today are near-freezing, and infinitely bracing by comparison with those of the Mesozoic hothouse, when little of the sluggish ocean, shallow or deep, fell much below a tepid twenty degrees centigrade or so. The deep oceans began to stir, and mix in cold water. Many deepwater creatures, accustomed to the warm deep bath of the Mesozoic oceans, became extinct. Ice grew larger in Antarctica.

What had happened? Among the myriad changes taking place around the globe, suspicion has fallen on those that redirected the patterns of circulation of water in the oceans. A continuous seaway began to develop around the continent of Antarctica as it separated from South America, with the opening of what we now call the Drake passage, which sailors better know as the fabled, treacherous sea passage around Cape Horn. Meanwhile, around the equator, a circum-equatorial current was closed off as Africa and Asia slowly drifted towards each other, and as North and South America became joined. The Panama isthmus is so narrow that, in certain places, and on a clear day, one can see both Pacific and Atlantic oceans: but, however narrow, it still stops water from one ocean flowing into the other. Thus a cold current system—a giant cold-trapping thermos flask, if you like, was being created in the far south, while a warm-trapping thermos flask of equatorial currents was, more or less simultaneously, being dismantled.

There were more climatic ups and downs, each adjusting the tortuous connections between land, ocean, and atmosphere. And then came the first really cold snap. Two and a half million years ago, scattered pebbles and boulders began to appear in the layers of deep-sea ooze at the bottom of the northern Atlantic Ocean. They could only have been brought in by flotillas of drifting icebergs.

The causes of this final push into the icehouse are unclear. Rising mountains—the Himalayas in particular—have been implicated, perhaps because they altered patterns of airflow high in the atmosphere, perhaps because the vast new area of exposed rock, on being chemically weathered by wind and rain, sucked carbon dioxide out of the atmosphere, and removed a few panes from the global greenhouse. Warm currents drifted closer to North America, forming a snowgun of moist air. This nourished snowcaps on that continent, and so the great Laurentide ice sheet began to grow. By whatever means, the Ice Ages had begun.

The Ice Ages! It is difficult, now, to understand the perplexity and bafflement and sheer disbelief that greeted this idea, over a century and a half ago: the idea of vast walls of ice invading from the north to engulf entire landscapes. This seemed like science fiction, a Gothic fantasy on a par with a belief in dragons and fairies and industrious aliens that built canals on Mars.

Yet, it was obvious to the nineteenth-century scientists that something untoward had happened in the very recent past. This was a past recent enough to form the surface of the present landscape. It represented a history seemingly quite different from the gentle progression of shallow and deep seas, river floodplains, and desert dunes seen in the strata of the more ancient past. This was a geologically violent recent past, with gigantic boulders strewn chaotically over the landscape, and caked mixtures of sand and mud and pebbles everywhere, and gravels seemingly laid down by torrential outbursts of water. All this now lay directly underneath the slumbering shires of middle England, and of rural Kansas and central Europe too.

There must have been a catastrophe. And the most likely of catastrophes here was a deluge, a Great Flood. With biblical Noachian overtones or without, this made sense. Floods, after all, were frequently witnessed, and so was the devastation they caused, as communities were buried by slurries of mud and pebbles, and smashed by the boulders dragged along by the raging waters. Few people, then, travelled to the far north, or to the glacier-draped mountains of the Alps.

A few inquisitive people did travel, though, and saw that the Alpine glaciers carried along with them a great deal of ground-up rubble. They saw, too, that masses of such rubble, now carpeted by grass and trees, were also scattered, miles down-valley from the present snouts of the glaciers. They surmised that the glaciers had, not long ago, extended farther than they do now. Some also saw a similarity between these masses of Alpine glacial rubble, and the boulder-rich clays and torrent gravels that covered most of the land surface of northern Europe. They put two and two together, and came up with a number entirely unacceptable to the geological establishment of the day. A Europe encased in ice remained as scientifically respectable as, say, a flat Earth.

Acceptance of the idea did come, but it came slowly. Politics, power, and influence played, as so often, a crucial role. One of the young lions of the scientific establishment, Louis Agassiz, who had made his name with his studies of fossil fish, went to the Alps intent on demolishing this pseudoscientific nonsense of invasions of ice. He came back a convert, and proselytized with passion and enthusiasm, risking both his reputation and his career.

He made powerful converts, among them the Reverend William Buckland, professor at Oxford, then perhaps England's best-known—and most eccentric—geologist, who reputedly ate his way through the entire animal kingdom available to him (bluebottle was the worst, he said, followed by mole), and wore his academic robe and top hat in field excursions with students. Buckland had been a sceptic, even when earlier shown the Alpine glaciers by Agassiz. Nevertheless, he invited Agassiz and fellow geologist Roderick Impey Murchison to the hills of Scotland and northern England, to examine the evidence to be found there. How do you distinguish a slurry formed by a torrential flood from a slurry smeared underneath an advancing ice sheet? Agassiz showed his companions a practical distinction, in that the slow crushing and shearing underneath a glacier can impart deep scratches and grooves to boulders and to the bedrock underneath. Not even the most violent flood will systematically produce such scratches.

Buckland was converted (though Murchison never was) and, also, began to spread the glacial gospel. In his turn he converted Charles Lyell to the idea, and this was crucial. Lyell more than anybody set out the philosophical framework for geology in the nineteenth century, not least with his idea of 'uniformitarianism', a grand word that means that past geology should be interpreted in terms of the long, slow action of processes to be observed at the present day. Indeed, his ideas were deeply influential well into the twentieth century, and remain relevant today. Lyell, soon after accepting the idea, was himself publishing on evidence of past ice in Scotland.

The idea of ice had come. Agassiz was told that he had 'made all the geologists glacier-mad here'. This was an exaggeration. It was to take another few decades for full acceptance, but glacial theory was no longer on the lunatic fringe. And very soon it became clear that the icehouse, once fully set in, was not at all simple. The Ice Ages are plural. The evidence for this was clear enough, once people starting poking among the glacial debris. For, here and there, encased in ice-lain deposits above and below, are little pond-fills of mud and silt, in which are encased fossilized leaves and twigs of oak and elm, and the bones of warm-loving animals such as the horse and the hippopotamus. The pattern was clear. Within the Ice Ages, ice had come and gone repeatedly.

It is now known that cold and warm succeeded each other with, geologically, bewildering speed. The Ice Ages were choreographed, quite precisely. The unmasking of the choreographer took over a century of study, during which time scientists chased whole shoals of red herrings. The identity of this choreographer is now becoming clearer, even as we, as a species, cheerfully throw sand into the gearbox of the intricate, only partly understood machine that governs the dance of the world's climate.

The tracking down of the choreographer of the Ice Ages began over a century ago, not long after the scientific community had, with various degrees of awe and reluctance, come to terms with the fact that much of northern Europe had, not so long ago, been entombed in thick ice, and then realized that the deep freeze had been interrupted, now and then, by conditions so warm that hippopotami and hyaenas roamed through what is now London's Trafalgar Square.

By the late-nineteenth century, James Croll, stubborn, eccentric, and Scottish, had learnt from astronomers of systematic variations in the Earth's orbit and spin. Thus, the Earth's orbit stretches into an ellipse and back again, while the Earth's spinning axis varies its angle of tilt, and this tilted axis also wobbles slowly. These irregularities of orbit and spin operate in mutually interfering cycles of 100 000 years, 40 000 years, and 20 000 years; the sum effect resembles not so much a metronome as an inventive jazz drummer, playing variations around a steady time beat. Surely, Croll reasoned, these irregularities would cause systematic changes in the amount of sunlight the Earth receives and in its seasonal balance—that is, whether winters or summer in one hemisphere or another would be relatively warm or cold. And that, he reasoned, might have driven the waxing and waning of ice sheets during the geologically recent Ice Ages. The scientific establishment looked on, intrigued but unconvinced.

The Serbian astronomer Milutin Milankovitch, in the early part of the twentieth century, spent a lifetime developing these ideas with mathematical rigour (painstakingly effecting the calculations without the benefit of computers) and vigorously championing them. Initially, some people listened. But these were ideas ahead of their time, much like Alfred Wegener's lone championing of the seemingly crazy idea of continental drift a few years later, and they suffered much the same fate. Most geologists pooh-poohed—and then simply forgot—the idea that minute astronomical changes might control the coming and going of vast ice sheets. They had the best of reasons for their scepticism.

The tiny changes in sunlight seemed a pretty flimsy influence on the formation of rock strata, when compared with, say, earthquakes, volcanoes, and the growth and decay of mountain belts. And it was proving exceedingly difficult to work out the pattern of the coming and going of ice, based upon the evidence of climatic change preserved on land. For the terrestrial record was exceedingly patchy, as any new influx of ice tended to bulldoze and destroy the evidence of past glaciations. In any event, geologists were beginning to cope with dating events in Earth history to the nearest few million years: distinguishing climatic changes separated by only a few tens of thousands of years was another scale of problem altogether.

Besides, the many studies of terrestrial Ice Age deposits seemed, by the middle of the twentieth century, to be coming down firmly against the ideas of Milankovitch. On every continent, it seemed that, over the past million years or two, ice had come and gone about four times. This timing did not appear to fit at all with the astronomical predictions. The ideas of Croll and Milankovitch seemed to be dead and buried.

These objections seemed perfectly sensible—but, astonishingly, were wrong. From the 1970s, scientists, notably Cesare Emiliani, John Hays, John Imbrie, and Nick Shackleton, in trying to understand the Earth's Ice Ages, had turned away from the tangled confusion of the evidence on land, where the grinding ice sheets had bulldozed away almost as much geology as they created, and looked into the stillness of the deep ocean floors. There, delicately manipulated drill-strings probed and sampled the deep-sea oozes—and revealed a story that, virtually overnight, changed our concept of the controls on climate history.

The story is now a classic. These researchers focused on a seemingly unlikely facet of these oozes: on whether the oxygen atoms in the limy shells of entombed and fossilized micro-organisms were 'heavy' (containing more of the stable isotope of oxygen of atomic weight 18) or 'light' (normal oxygen, atomic weight 16). The proportion of these in the fossil shells varied systematically from layer to layer. And this variation traced out a pattern in time that, once statistically interrogated, rang out periodicities of 100 000, 40 000, and 20 000 years—exactly those which had been predicted by astronomical theory. When Hays, Imbrie, and Shackleton published this in 1976, their predecessors Croll and, especially, Milankovitch received not just vindication, but well-nigh scientific canonization.

This isotopic pattern was produced by the geologically recent partnership of huge continental ice sheets acting in tandem with the oceans, to make the Earth's surface into one giant fractionation column. When the ice sheets expanded, they preferentially sucked the easily evaporated light oxygen-bearing water molecules out of the oceans, which were left isotopically 'heavy'. When the ice receded, the light oxygen flooded back into the oceans. And generations of micro-organisms, unconsciously absorbing the oxygen into their shells, faithfully recorded the twists and turns of climate. This discovery gave us our blueprint for understanding past, present, and future climate, perhaps the most pressing question our current civilization has to understand and resolve.

The drill-cores taken from deep-sea oozes revealed, unambiguously and repeatedly, from core after core, that the pacemaker driving the myriad climatic changes of the Ice Ages was essentially astronomical. Yet, the slight variations in the amount and distribution of sunlight falling on the Earth seem wholly insufficient to produce such a drastic climatic effect. But produce them they did. Why?

The astronomical pacemaker seems to hold terrestrial climate in a tight grip by hiring some muscular henchmen, who have followed its promptings with little hesitation. Chief among these enforcers are the greenhouse gases, mainly carbon dioxide, and also methane. The evidence lies in bubbles of fossilized air, trapped in up to a million years' worth of compressed annual snow layers in the great ice caps of Antarctica and Greenland. From these, the concentration of carbon dioxide and methane in ancient air can be measured directly. These gases have faithfully replicated the astronomical patterns. Quite why is not entirely clear. But whatever the precise mechanics, astronomy, greenhouse gases, and climate have so far acted in virtual lockstep. At certain astronomically defined thresholds, global temperature, sea level, and the levels of greenhouse gases rose and fell in such close harmony that cause and effect are difficult to disentangle.

The overall pattern is undoubtedly that which Milankovitch predicted. But, there are a few ways in which the observed pattern of climate change departs from that predicted from the astronomical predictions. First, the predicted, theoretical pattern shows a complex mixture of the three main pulses, deriving from regular variations in the tilt of the Earth's axis, in its slow wobble, and in the stretching of the Earth's orbit from a circle into an ellipse and back again. The observed record of climate indicates, though, that one of these pulses was generally dominant in any given interval of time. In the last half-million years or so, it has been the 100 000-year-long pulse of orbital stretch. This dominance would not have been predicted from the astronomical calculations: the 100 000-year cycle should now be subdued, but it is not. Evidently, there is something at the Earth's surface that has acted to modulate, to resonate with, just one of these astronomical periodicities at any one time. Some suspicions are focused upon the great, sluggish ice caps of the high latitudes which, perhaps for any given size and stage of development, may only happily dance to one of the available tunes.

Secondly, the predicted astronomical curve is complex, but overall, each peak of warm and cold is symmetrical, predicting equal rates of warming and freezing as the warm and cold periods succeeded each other. It was not so in reality. The cores of ooze and ice show a consistent, strongly asymmetrical sawtooth pattern, indicating that the Earth's Ice Age history showed a strong pattern of slow cooling and abrupt warming. Thus, the Earth has tended to slide slowly into a glacial phase over some tens of thousands of years. But, when it has come out of a glaciation into a warm, interglacial phase, it has generally done so much more quickly, amid collapsing ice sheets and quite extraordinary bursts of warming and of change in sea level that, overall, could rise globally by more than a hundred metres as global cold gave way to global warmth.

Within individual Milankovitch cycles, too, particularly during cold phases, there have been abrupt swings in temperature, the most closely spaced being every thousand years or so. These do not link easily with orbital changes, and remain mysterious. The great ice caps are implicated in such fine-scale distortions of the Milankovitch pacemaker. They are best documented over the last few tens of millennia, over the heyday, then decline and fall of the northern ice sheets. Here, there were a series of closely spaced, dramatic collapses of ice sheets (Heinrich events, these collapses are termed), expressed as distinctive, far-flung layers of cobbles and boulders buried on the sea floor of the north Atlantic. These layers of debris mark the passage of colossal flurries of jostling icebergs. Where were these icy armadas launched from? The rubble layers can be tracked back to the eastern coast of North America, and the disintegration of the great Laurentide ice sheets which, lubricated by outpourings of meltwater, lost their grip on the land, and literally slid into the sea.

Each of these surges of icy rubble into the Atlantic was big enough to raise sea level, by up to several metres at a time. The transition seems to have been disconcertingly abrupt, an abruptness that could—and likely did—register on a human timescale: the peoples then living by coastlines would almost certainly have had to run for the hills. Literally.

So, how will sea level behave in the immediate future? This is currently a question of some concern to us, both personally and collectively. Let us, though, consider how it will affect our standing as objects of future curiosity from an ethno-archaeological point of view. The likely behaviour of sea level in a near future measured in centuries and millennia will, very likely, give the archaeologists of the future an extraordinarily good view—if not necessarily an extraordinarily good understanding—of the ruins of the human empire.

All things being equal, the Earth should now be sliding slowly back towards the grip of ice. We have had almost exactly 10 000 years of warmth, which is already longer than any other warm spell over at least the last 400 000 years. This climatic gift has allowed our civilization to develop to its current extent, where it colonizes virtually every agriculturally productive part of the globe.

It is becoming increasingly clear that all things are not equal, even if the exact state of this inequality is currently provoking fierce debate. The control buttons of the Ice Ages are still puzzling to us, even as we are collectively pushing a good number of them, and wondering how the autopilot is doing. The autopilot, mind you, has been functioning very steadily up until now, thank you very much, but it is a mite worrying that we still do not exactly know how it does this trick. Is a rapid sea level rise likely to happen? We are in unknown territory here. We are already in a warm phase of the Earth's Ice Age climate. So, will sea level go up at all, let alone go up with terrifying and civilization-threatening speed?

One might take a sceptical stance on this. We are accustomed to thinking of the world as a pretty stable place, an unchanging natural backcloth too vast to be affected by farming, by real estate development, by the normal functioning of a growing economy. These lie at the heart of all that maintains our livelihoods. But, the world of naturally growing industrial economies may have a greater effect on the Earth than we might imagine. For instance, levels of carbon dioxide—the most important greenhouse gas—are rising dramatically. Since about 1950, levels have been directly measured in the atmosphere. Levels are now going up by almost half a per cent a year (a rate of increase that is now accelerating). The fossilized air bubbles trapped in the world's ice caps show that today's levels of carbon dioxide are now about 30 per cent higher than at any time in the past million years.

To be more exact, carbon dioxide levels are now about 380 p.p.m. (parts per million, by volume) in the atmosphere, and rising now by about 1.5 p.p.m. every year. The ice cores show that pre-industrial levels in this and in preceding interglacial phases were about 280 p.p.m.; this is the 'natural' level in those climate states. In glacial phases, carbon dioxide levels go down to 180 p.p.m. Thus, as regards this gas, humanity has changed the Earth's atmosphere by an amount equivalent to that between fully glacial and fully interglacial states.

These bubbles of fossilized air also tell us how much of another potent greenhouse gas, methane, there was in pre-human air. Methane is naturally produced from rotting vegetation and such like, but human civilization does considerably better than raw unadorned nature, though by admittedly primitive measures, such as by cultivating many cows which collectively pass a lot of wind. Levels of methane in the atmosphere are now more than twice as high as at any time in the past million years.

There is currently a good deal of discussion about what all of this might mean. But the most widely accepted scenario, among the scientists working on this phenomenon, is that climate will quickly warm to temperatures not seen for the past 10 million years or more. 'Quickly' means in a few hundred years: geologically that is virtually instantaneous.

But hold on just a minute—a sceptic might say—*there are problems with that prediction*. Indeed there are, for we are going into the unknown as humanity unleashes its experiment with global climate, and some of our assumptions might be wrong. For a start, the ice cores have shown us that temperature and greenhouse gas levels act in such close lockstep that it is difficult to disentangle cause from effect. Could it be that carbon dioxide levels in the atmosphere went up during each interglacial *because* the climate warmed? And, we are already at the peak of a warm phase of the Ice Ages—could it be that we are at a temperature ceiling, which a modest change in CO_2 concentrations will not shift? And could a modest amount of warming put off the day—surely soon due—when the world inexorably slides back into the next Ice Age?

Let us look back a little, first. Not far, just half a million years or so. But it has been a crowded, eventful half-million years. It has been much studied of late, and some things are becoming clearer. In this half-million years, the dominant influence on climate has been the 100 000-year cycle of the variable 'stretch' of the Earth's orbit around the sun. So, very roughly, cold glacial phases roughly 100 000 years long have alternated with warm interglacial phases roughly 10 000 years long (or about half a 'wobble' cycle), and we happen to live in the latest of these, indeed we are about ten thousand years into it . . .

Well, that is the cartoon view. But what really happened? Here, those amazing ice cores have provided the devil in the detail. And they show that each of the interglacial phases was individual. The previous interglacial phase comparable to ours was some 125 000 years ago. It was not quite 10 000 years long, but it was just a shade warmer than now. Mostly less than a degree, but some 5 degrees warmer over Greenland, where, doubtless because of this, the ice cap seems to have been significantly smaller. If there was less ice, then there was nowhere for the water that made it up to go except, of course, into the sea.

Now it is ferociously difficult to work out exact levels of past sea level, because the land surface can also rise and fall (not least when it is pressed down by the weight of enormous ice caps). But some more or less stable tropical islands (which have never been covered by large ice caps) have remnants of cliff lines, some 125 000 years old, which are about 5 metres above present sea level. Five metres is trivial, geologically speaking. Sea level has been much, much higher and much, much lower in the past. But raise the sea 5 metres above its present level, and it would submerge, for us land-dwellers, a quite inconvenient amount of the landscape

There is another point of view, of course. This one says that our current warm period started some ten thousand years ago. Ergo, we are just about to slide into another glaciation. Thus . . . thank goodness for global warming!—it has put off the evil day when we have to cope with Siberian climates in Maine and in Manchester.

This idea of the metronomic regularity of the ice ages lasted a long time, sustained by, perhaps, an over-simplified view of the climate's astronomical pacemaker, so painstakingly calculated by Milutin Milankovitch. Cracks in this idea began to appear, when an international team of scientists drilled through four hundred thousand years' worth of ice at the Vostok research station on Antarctica. The record of climate shown in the borehole—including bubbles of fossilized air—showed that the three warm periods prior to ours lasted little more than seven thousand years before the global chill set in again. So we were in an unusually long period of stable warmth—that, incidentally, had allowed our civilization to flourish; but, we seemed to be living on borrowed time.

It is a pity the borehole did not penetrate a little deeper. For, round about that time, a deal of attention was being paid to the fourth interglacial period prior to ours, or Oxygen Isotope Stage 11 as it is known more technically. Stage 11 was preserved as a few fossilized patches of pond- and estuary-mud on land, and as layers of ooze on the deep sea floor. Some of those estuaries had seemed to be perched a bit high inland—but then, it was a long time ago, and the Earth's crust has gone up and down, particularly when pressed on by the weight of millions of tons of ice.

Stage 11 appeared in a different light, though, when the most complete layers of oceanic ooze were studied, and when evidence of a sea level perhaps up to twenty metres higher than now was seen on tropical oceanic islands far from the crust-deforming influence of the ice sheets. It seemed to have lasted for longer than ten thousand years—not just by a little, but by all of twenty thousand years. Stage 11 appeared thus, to be thirty thousand years long, and warm enough to have melted parts of both the Greenland and Antarctic ice sheets. (The extraordinary duration of this warm phase has since been confirmed by the latest, and deepest, borehole through the Antarctic ice, all of 750 000 years' worth, which sampled Stage 11 in its entirety.)

Worse (or better, depending how long-term your point of view is), Stage 11 seems to be, astronomically, the nearest equivalent to our own current interglacial—which therefore might be naturally programmed to run for thirty thousand years, and become warm enough, in the near-ish future, to bring the sea flooding in. As some of the scientists who carried out these studies mildly remarked, the human-induced greenhouse effect cannot help but encourage this process. That might be an understatement.

These examples from the Ice Ages simply represent minor natural fluctuations in climate and sea level between successive warm phases, without any major, sudden increases in greenhouse gas levels. The differences in climate and sea level they produced were geologically trivial, but of a scale which, if repeated in the near future, would put human civilization into considerable difficulties. So what about our very own home-grown X factor, the burning of billions of tons of fossil fuels every year? Do we have any geological precedents for this?

We do, but we have to look much further back in geological time to look for examples of sudden, enormous outbursts of greenhouse gases into the atmosphere. At least two have been identified, in the Toarcian Age of the Jurassic Period, some 180 million years ago, and in the early Eocene Epoch, around 55 million years ago.

The world then, in both instances, was already very warm, with little polar ice. Nevertheless, extraordinary things happened. Sudden changes in the carbon isotope composition of fossils from both land and sea show that greenhouse gases flooded into the atmosphere. The isotopes themselves do not say whether mainly carbon dioxide or methane was involved, but plausible scenarios suggest the involvement of both (say, by deriving carbon dioxide from massive, geologically rare volcanic outbursts, providing initial warming which in turn destabilized methane that had been stored in permafrost or in ocean floor sediments). Whatever the precise mix or source of gases, the amount of warming is now broadly established, again from isotope ratios preserved in fossils. Rapid warmings of the order of between 5 and 10 degrees centigrade took place globally, the temperatures declining back to background values over several tens of thousands of years, probably as the excess greenhouse gases were slowly drawn out of the atmosphere by chemical reactions associated with rock weathering.

These ancient events strongly reinforce the idea that global warming will take place over the coming centuries, especially as the rates of greenhouse gas release then are thought to have been comparable to (but much slower than) the seven billion tons of carbon that we currently release annually into the atmosphere by burning fossil fuels. Such temperature surges show the Earth behaving in a non-linear fashion when reacting to environmental stress: that is, it tends to 'flip' from one more or less stable state to another, and this kind of behaviour is inherently difficult to model or to predict. So, sudden increases in both temperature and sea level are likely—but we cannot say exactly when. There will be, as the oceanographer Wallace Broecker has said, unpleasant surprises in the greenhouse.

So, the greenhouse gases, the enforcers of the Milankovitch edict, now have a different godfather. This is we, the people. The likely result of this power transfer—global warming—looks set to disconcert us all. Some consolation for our earthly struggles as the waters rise around us may or may not prove of comfort: our chances of permanent fossilization are going to be enhanced immeasurably. If we and our children are very unlucky over the next few decades, and the waters rise swiftly, then many of our cities may be as well preserved as Pompeii, as though in aspic.

At this point, there are two main points to muse on. First, how quickly is sea level going to rise as our gas guzzlers do their stuff? Secondly, are we creating just a temporary blip before the astronomical pacemaker reasserts itself and ice once more takes a grip on the planet; or, are we dismantling the entire mechanism of the Ice Ages, and plunging ourselves back into the hot climate enjoyed by the dinosaurs for tens of millions of years? If the latter is to be the case, then our geological immortality would be certain. As a species, we are extracting a few hundred million years' worth of buried carbon from the ground and launching it into the atmosphere as carbon dioxide in little more than a century. This is providing a mode of warming that is geologically unique. If, by our global chemical experiments, we provoke a few ice sheets into collapsing, the sea level changes will likely be difficult to deal with, as living space needs to be found for several hundred million displaced souls. The fossilization of our human empire will, though, have been made immeasurably easier. As sea level changes, the edge of the sea will move into a new location.

There are matters of historical consequence here. The edge of the sea is not just a thing of pleasure and wonderment and long walks along the beach by the evening sun. It is a thin line where energy is exchanged: where wind energy, gathered and transmitted by the waves over thousands of square kilometres of sea, is concentrated and converted into the pounding turbulence of the surf zone. This energy, unleashed, bites away at the land to create cliff-lines and the trillions of beautifully shaped, wonderfully segregated sand grains of the world's beaches.

Let us consider, thus, the importance of the *rate* of sea level rise. When the sea floods a landscape, it can do it in two ways. It can chew it up, by the action of pounding waves at a shoreline over a protracted period—or it can simply drown it. The chewing-up is brutally efficient: the sea captures energy over its entire surface, as winds and storms stir its surface into ever-larger waves. That energy has nowhere to go except to be explosively dissipated in dismantling the first land surface that the waves get to. So, given enough time, the sea will simply cut through any adjacent landscape, to the level of a wave-cut platform at the base of a cliff-line. Many geologically ancient sea level rises (or to give them their technical term, marine transgressions) have done this, and can be seen in rock strata as flat erosion surfaces, above which the land mass has been pounded into pebbles, sand, and mud, and then washed away, ultimately to form rock strata in distant lands.

Thus, an advancing sea is a double-edged sword, at least as regards the preservation of vast Earthly empires for future civilizations to wonder at. Let it creep in, inch by inch, and it will grind away, slowly but effectively, at our monuments of concrete and steel. Let it bite away at our present landscape for long enough, and it will level our cities to their foundations. Skyscrapers and semi-detached houses alike, roads and railway lines, will be reduced to sand and pebbles, and strewn as glistening and barely recognizable relics along the shoreline of the future.

But, if the sea rises quickly enough, and there is not time for the waves to do their work, landscapes may be drowned entire. Only a few metres beneath sea level, and what was the land now lies below the destructive surf zone. A hundred metres below sea level, and even the most violent storm waves can scarcely be felt. So, let the sea flood in, with its level jumping by metres over centuries or decades—or perhaps even years—and there simply will not be time for this wave energy to erode the landscape. The shoreline will literally skip over the low-lying coastal zone as it finds a new level, say five metres higher, and, once stabilized there, it will began to erode again, to create a new cliff-line. And much of the coastal zone that has been skipped over will be spared the erosion, and its contours will then be covered with layers of marine sand and mud.

The process might be beginning. Sea level globally, over the past century, has been creeping upwards at the rate of about two centimetres each decade. Part of this sea level rise has been due to the thermal expansion of the ocean waters as the Earth has gradually warmed over the past century. Another part results from melting glaciers and ice caps.

Whichever is the case, the latest news from the Earth's polar regions and high mountains suggests that the process is beginning to accelerate. Even at today's only slightly elevated temperatures, following a rise of around half a degree centigrade, mountain glaciers are receding significantly, as also seem to be, locally, the margins of the ice in Greenland and Antarctica. The Greenland ice cap is vulnerable, not least through its very height, which means it creates its own climate. The ice surface now reaches the chilly height of 3 200 metres above sea level. Lowering this surface through initial melting and ice loss would take it into lower—and therefore warmer—regions, which would increase the rate of ice melt, and so lower the ice surface still further . . . and so on. It is a kind of positive feedback that can threaten the entire mass of ice. Once the ice has entirely melted, its loss would be effectively permanent—and would raise sea level globally by about six metres.

In Antarctica, some large ice shelves, such as the one named the Larsen B, have suddenly broken up in the past few years. This, and the melting of the released ice masses, cannot in themselves cause sea level to rise, as floating ice already displaces its weight in water (try watching an ice cube melting in a glass of water and see if you can see the level rise). The very big question here is whether this break-up has encouraged increased streaming of ice from much farther inland, which potentially represents the beginning of a phase of much more serious ice-sheet collapse. At the time of writing, the most recent observations suggest that it has.

Increased warmth may also, though, provide a counterbalance. If this extra warmth increases the evaporation of water from the sea surface, some of the extra water vapour can drift over an ice cap, and then condense to produce extra snowfall—the snowgun effect—and that extra snowfall will ultimately lead to thicker ice. In recent years, it has been possible to monitor this, by measuring the height of the Greenland and Antarctica ice caps very precisely, using radar from satellites. Currently, the interior of the Greenland ice is growing, while the margins are melting. Likewise, the interior of the great mass of ice that is East Antarctica is growing, while parts of the ice mass on the West Antarctica peninsula—including the Larsen ice shelf—are shrinking.

This type of behaviour has shown clearly that ice sheets are not relatively inert masses responding sluggishly to temperature change. Rather they are now perceived in much more dynamic terms, responding disproportionately to small changes in external conditions. But which tendency will dominate? Will increased snow accumulation offset or even outweigh increased melting, causing sea levels to fall?

The progression of events is likely to be complex, and the effects of global warming will likely reverberate unsteadily through the Earth's climate system for centuries and millennia, like a rogue billiard ball bouncing through one of Heath Robinson's more complicated devices, haphazardly redistributing the pattern of levers.

Geological evidence suggests that global warming episodes are typically linked with sea level rise. Such rises may proceed in giant steps, as ice sheets collapsed in the past to release 'iceberg armadas', with sea level rises of a few metres, rather than a few tens of centimetres, in a century. We may not quite be at that stage yet, but to keep a very close watch on the world's ice caps would seem sensible.

If sea level continues to inch upwards, what would its effect be? London, for instance, is centred on a low-lying river just upstream of a strongly tidal estuary. This city is still mostly untroubled, but is vulnerable to flooding if low atmospheric pressure sucks water upwards while strong winds drive it shoreward during a high spring tide. It is for that precise combination of circumstances that the Thames barrier was built. The sea level rise now predicted for SE England is some six centimetres per decade, a higher figure than the global average because the land is sinking while sea level is rising. If a hotter, more energetic climate unleashes extreme flood events more frequently, and if sea level rises faster than predicted because of accelerated polar ice melt, then the Thames barrier—or its future successor—could be overwhelmed later this century.

Given such a trajectory, central London can expect to be flooded, suddenly, dramatically, during a major storm. Far-fetched? Well, New Orleans has already experienced it. The first such flood will cause an almighty mess (imagine getting the water out of the underground system). That is likely to be cleaned out, at enormous expense. Perhaps even a second, or even third, flood will be weathered. But eventually that part of London will simply be abandoned—as will Amsterdam also, for its sea defences will likely give way about the same time. This is not inevitable, but it seems a geologically reasonable prediction.

So, a century from now (two, if we are lucky), central London, Amsterdam, Venice, New Orleans will be abandoned and human civilization (assuming pandemics and wars have been survived) will be relocating towards higher ground. A good deal of Bangladesh, Florida, the South East Asia coastal plain, the Nile delta—the list can stretch on and on, for low-lying coastal plains have been comfortable and productive places for humans to settle in—will be under water.

There is much food for thought there, but we will content ourselves with sticking to our brief. Thus, a lot of real estate will be under water, and beginning to be covered with sand and mud. This would immediately put these regions beyond the reach of erosion—except perhaps for a little localized scouring by strong tidal currents—and into the kingdom of sedimentation. Our drowned cities and farms, highways and towns, would begin to be covered with sand, silt, and mud, and take the first steps towards becoming geology. The process of fossilization will begin.

For our explorers, though, there is a long way to go to discover—and interpret—what might remain of these drowned cities. Many questions remain. How can cities be petrified, for instance? Where will they go, as the continents drift on? But more importantly—how can they be found? Remember that the most ancient of these represent only a few millennia. Most of them (all the North American cities, for instance) are only a few centuries old. That is not much more than a millionth of the time span separating our epoch from that of our future explorers, and it is of the order of a hundred millionth of all of Earth history. That is a small needle in a large haystack.

We will not be unearthed by chance. Our explorers will need some guidance. They will need to be led to the petrified cities by following a trail of evidence that humanity is dispersing more widely in the strata forming today, just as a detective would follow clues towards the scene of a crime. And for that, of course, the detective must develop a methodology to allow them to read these clues and put them into order and context. To be able to read the signals, therefore, these visitors from a distant star system will need to develop some particular skills. They will have to become Terran geologists.

Dynasties

The historical detail is almost overwhelming: quite as complex—and as time-consuming for our scientists—as the almost limitless biological variety. We are finding ever more petrified remains, and have to improvise constantly and to develop new types of science, for much of the material we uncover is stubbornly resistant to the standard analytical techniques. We may be here some time yet, for there is much to do; our research seems to have barely begun.

FUTURE EVIDENCE

Developing a methodology is everything in a science. Once you have it, you can go on to extract information, facts—a narrative—from the natural world. To human scientists and non-scientists alike, the use of fossils as evidence of past events on Earth is now taken for granted, is indeed ingrained into popular culture. Dinosaurs, for instance, stalk through our TV screens and cinemas and shopping malls, as virtual animations and plastic models and soft fluffy toys and comic book covers. An Age of Dinosaurs is widely accepted as a long-vanished era, a world lost within deep time.

Our extraterrestrial investigators will, at some stage in their studies, be ready to try to recreate for themselves the eras of long-vanished animal and plant dynasties on this planet, to construct a coherent history out of the scattered relics preserved in the Earth's abundant strata. By coming to understand the Earth's marvellously regulated heat-release engine, that drives the tectonic plates, they will appreciate the continuous creation and preservation of strata. By getting to grips with the more subtle puzzle of how sea level has risen and fallen, they will have some idea of the finer controls on the preservation of the stratal record. And, as they grapple with these problems, they would undoubtedly try to put the strata themselves into some sort of order, just as did our Victorian and pre-Victorian predecessors. These pioneering geologists, after all, could recognize a prehistory when they saw one, even as they were still far from divining the workings of the Earth machine that lay at the heart of the story they were pursuing.

What kind of strata will be available for study, one hundred million years from now? Many, if not all, of the classic fossil localities that we treasure today will have gone forever, eroded into scattered grains of sedimentary detritus that will ultimately accumulate on sea floors of the future. The Solnhofen Limestone of Germany, that yielded the archaeopteryx, will likely be gone. The Burgess Shale of British Columbia, with its wonderful array of early soft-bodied organisms from the Cambrian Period, half a billion years back, is almost certain to disappear, perched as it is high up a fast-eroding mountainside.

Then there are the strata now buried, that are or will be on an upwardly bound tectonic escalator. This will make them, one hundred million years from now, arrive at the land surface of the far future, available for inspection and examination. We have a slight inkling of what might be included here, as our species has drilled widely through into deep-lying strata around the world for science and (more often) for profit. Perhaps the meteorite crater implicated in the extinction of the dinosaurs, at present buried over a kilometre down beneath the Yucatan peninsula of Mexico, will by then be exposed at the surface. Or there are some fine fossil-bearing mudrocks of my own acquaintance, now several hundred metres under the English Midlands, borehole core samples of which are currently archived at the British Geological Survey; I would quite like to borrow a little immortality to be able to see more of those rocks, when they eventually arrive at the surface. And there will be strata in abundance, yet un-drilled and unseen; all of this will contain information on times past, and a few will bear treasures as rich as the Solnhofen and the Burgess.

Then there will be future strata, as yet undeposited. That is, all the layers of sediment that will accumulate on the Earth's surface in the next hundred million years. Much of this, one hundred million years from now, will be buried deep below the surface. A good deal will have been buried and then eroded again and so transformed into yet newer strata. A reasonable amount, though, of this next hundred million years' worth of strata will be at or near the future land surface, just as today rocks representing the last hundred million years (back to the early Cretaceous Period, that is) are widespread and easily accessible.

And then, separating the ancient, yet-to-be-exposed strata and the newer succession that is yet to be deposited, will be the strata of the present, of the geological instant that is *now*: the interval of human civilization. It is less than ten thousand years in duration, this interval, and so just one per cent of one million years, which is itself just the small change of geological time. Seemingly with the stratigraphic thickness of a piece of cigarette paper, it will be caught between the ancient past and the substantial future. Yet, I would hazard a guess that our future investigators will discover it. And marvel.

We cannot say what strata will be present or where as our future archivists start their work, but we can make a stab at the rough proportions of the stratal terrain that will be available for them to study. We can look today at the outcrop of all the solid rocks on the Earth's land surface, those at the surface now and those that lie just beneath a thin cover of the loose sands and gravels and clays of very recent 'drift' deposits. If we take the map of the British Isles as exemplar, then some 20 per cent of the land area is underlain by rocks that are a hundred million years old or less. Another 70 per cent or so is made up of rocks between a hundred million and five hundred million years old. About 10 per cent is made up of the older rocks, the Precambrian rocks, that extend back to more than three billion years before the present.

By no means all of these rocks are stratified sedimentary rocks, and so not all yield a clear signal of past environmental conditions at the Earth's surface. There are plutons of granite and gabbro, coarse-grained igneous rocks that crystallized far beneath the surface and then were carried slowly to the surface as the overlying rocks were eroded away. There are rocks, both sedimentary and igneous, that have been recrystallized at high temperatures and pressures in the roots of mountain belts. By and large, the older a rock is, the greater are the chances that it has been involved in such a major tectonic traffic accident, and so unmetamorphosed examples of the most ancient rocks on Earth are rare. Even metamorphic rocks, though, if not heated too strongly or compressed too tightly, still bear imprints of their original character, and may even yield fossils. Once they are thoroughly heated and recrystallized, though, this original story is obliterated, and is replaced by one that charts the slowly evolving world deep beneath the surface. This is Pluto's realm, and it is dominated by changing regimes of heat and pressure and slow fluxes of subterranean fluids. If one could dig a tunnel down to it, and somehow prevent the walls from caving in, then it would be a world devoid of oxygen, a world soon devoid of life (past the first couple of kilometres, beyond which even the most resilient of microbes can no longer persist), a world of darkness, and, as, as temperatures rise on descending ever deeper, of oven-heat. Most of the rocks we would see have been somewhere in that realm for far longer than they have lain at the surface.

A hundred million years from today, then, roughly similar proportions of rocks of different ages, variously stratified and non-stratified, will be present at the Earth's surface. There will be proportionately less of the rocks we see today, by the amount that they will be covered by younger rocks (those beginning to form today) and also by the amount that more of the continents may lie submerged under the (likely) higher sea level of the future. Beneath those oceans, the steady progress of plate tectonics, of the creation and destruction of oceans, will have been taking place. A little more than half of the ocean floors of today will have been destroyed, and replaced by new ocean floors. This will obliterate a good deal of the evidence now available to us, and replace it with new evidence, of present and future times. (An interval within the slowly accumulating ocean floor oozes just a few centimetres thick will be all, in this realm, that represents the human empire.) All in all, there will be plenty to study: a cornucopia of

Earth history going back three billion years and more. It will not be ignored in any future exploration of this planet.

The major episodes of this history will still be clearly readable. Even if all of the evidence at our current disposal will have been effaced, it will have been replaced by new evidence brought to the surface by crustal uplift and erosion; this new evidence may show some aspects of Earth history more clearly than is afforded at present, and some more obscurely. Much of the evidence, though, will simply represent lateral extensions (or what geologists term correlatives) of the same strata that we now see at the surface. Thus, on whatever will remain then of Britain as an island, Chalk rock may still be exposed, but it will be the Chalk rock that now lies at depth, hundreds of metres beneath London, say, or beneath the northern part of the Isle of Wight. It will be essentially the same rock recording the same time interval and the same sea floor, but it will represent different parts of that sea floor. Thus, different insights will be available from the strata that will then be at the surface. The bones of Earth history, though, are unlikely to change.

EVENTS IN TIME

The very oldest history—indeed the bulk of Earth history—will still be represented, one hundred million years from now, by the Precambrian rocks that form the ancient cores of the continents. These may be time- and weather-worn but, as a whole, they are virtually indestructible. They have been around for long enough to have been involved in sundry continental collisions, and so are mostly crumpled, recrystallized, metamorphosed. Nevertheless, sufficient recognizable sedimentary strata remain for a telling absence to emerge, an absence that was already noted, in our time, by Darwin and his contemporaries. There are no fossils in these rocks, until their very youngest parts are reached—or, at least, no obvious impressions or casts or moulds of any of the major groups of Earthly life: no arthropods, no molluscs, no leaves or twigs, no fish, no starfish or sea-urchins, no corals.

And so the first major natural division of the Earth's strata will emerge, once our future investigators begin to get their eye in: a succession of rocks without obvious fossils below and a succession of rocks above that often abound in fossils (and within which even most 'unfossiliferous' strata will yield recognizable fossils, if one looks hard enough). This will not by any stretch of the imagination be an interpretative construct, a boundary to be imposed arbitrarily within a mass of strata so as to form a formal subdivision within them. Rather, it is now—and will be a hundred million years hence—a clear natural division that further study can only further anatomize and characterize, a boundary whose sharpness will be enhanced rather than blurred with more information. The Earth was devoid of familiar multicellular life-forms for part of its history, and then, rather abruptly, acquired them.

Many questions can be asked of this revolution in the Earth, an origin and radiation of multicellular life that we call the Cambrian explosion. How did it happen? Was it triggered by biology, or by some change in the Earth's environment? Just how abrupt was it? But first there is another, more fundamental question. *When* did it happen?

The geological vision of the Earth's past is essentially a history based on a piecing together of evidence preserved in rock strata, where everything boils down to saying whether any one event—the eruption of a supervolcano, say—took place before, during, or after some other event, which might be, say, the appearance of complex life at the beginning of the Cambrian Period. This biological event, marked by the appearance of abundant fossils within strata, itself forms a marker that can be traced within strata, even within widely separated and tectonically fragmented strata, all around the world. It is an anchoring point in Earth history, to which other events worldwide—other super-eruptions, say, or meteorite impacts, or submarine landslides, or the migration of a particular train of sand dunes—can be related in time, based on whether those events are preserved in strata that lie below, above, or within the same strata that yield these first obvious fossils.

This would be the beginnings of an ordered succession of geological events. But it is a relative history, not one that is measured in years. We might have no idea whether any particular event recorded in strata happened ten thousand years ago or a trillion years ago, but we would suspect that the true date lay somewhere between the two. This was, indeed, the situation in which Victorian geologists found themselves, in which they had pretty well sketched out the grand outlines of Earth history, without having much of a means of working out quite when the amazing succession of events that they were uncovering actually happened. After all, a fossil trilobite—one of the groups of fossils to abound after the Cambrian explosion—did not come complete with a sticker saying quite how many years ago it turned from being a living creature on a Cambrian sea floor into a stony relic.

It still does not. But what the Victorian geologists did not suspect was that the volcanic ash that might have rained down on that unfortunate arthropod *did* come complete with built-in date stamps which, if properly read, can tell us how long ago the eruption took place. These date stamps only became readable when the phenomenon of radioactivity was discovered, and then exploited for the radiometric dating of rocks and minerals. This is now basic physics and chemistry. An extraterrestrial civilization technologically advanced enough to visit the Earth, and curious enough to investigate its history, will have already started its radiometric dating programme. It will want to know how old the Earth is, and just when it acquired its remarkable array of living organisms.

ATOMIC TIME

The technique is straightforward enough. Some types of atom, such as uranium, break down at a steady rate into other types of atom (lead—ultimately—in the case of uranium), the rate of breakdown being unaffected by fire, flood, storm, or extremes of temperature or pressure. Uranium is generally a rare element, but it occurs in some abundance (a few per cent) in a mineral called zircon, which is often found in, say, the volcanic ashes widely distributed over both land and sea in a super-eruption. In recently erupted zircon crystals, there will be significant amounts of uranium and virtually no lead, while in very old zircons a good deal of the uranium will have decayed into lead. Measure—very carefully, using a mass spectrometer—the ratio between the two, and one has a measure of the time of eruption. Work out—equally painstakingly—just how fast uranium decays into lead (this is a slow process: the decay rate is measured in millions of years) and one can calculate the number of years since the formation of the zircon mineral, which (one hopes) was forged in the magma chamber just before the eruption that hurled it on to the landscape. Each zircon crystal is, in fact, a very neat, self-contained atomic clock.

It is more complicated in practice, of course. There are in fact several types (isotopes) of uranium, each with different rates of breakdown into different isotopes of lead. Most lead is just common old lead that was always lead, and did not form from uranium at all (although, fortunately, the zircon crystals fastidiously exclude atoms of this type of lead from their mineral structure as they crystallize). And zircon the mineral is much, much tougher than old boots. This is mostly a good thing (the atoms in a zircon crystal are not easily reshuffled, and so this atomic clock is not easily reset to year zero by any of the convulsions of heat and pressure the subterranean world is prone to); but an old zircon crystal can quite happily find its way (across eons of geological time) into a new volcano, producing a rogue atomic clock capable of seriously confusing the unwary geologist.

Or an old, geologically adventuresome zircon crystal might be made of *two* atomic clocks, having a core of the original magmatic zircon and an outer layer of zircon recrystallized—and so with its atomic clock reset to zero—when caught, say, in the high temperature, high pressure roots of a growing mountain belt. Dating the entire zircon crystal will therefore give an average of these two distinct and time-separated events. One can today, almost miraculously, obtain *both* of the dates from a single compound zircon crystal, by aiming a finely focused beam of ions or of electrons separately at the unaltered core and at the reset rim, digging little pits some two-hundredths of a millimetre across within them, and knocking out enough atoms to enable the ratios of uranium to lead to be established from each.

So, a variety of pitfalls can be negotiated, with care and patience, and the adoption of a wary and sceptical demeanour on the part of the investigator. There are cross-checks that can be made. Uranium is not the only radioactive element that can be used for measuring the deepest terrestrial time (though its partnership with zircon is currently a much sought-after combination, because of the precision of the time estimates it can provide). There are radioactive forms of potassium, and of rubidium, and thorium, and also of lesser-known elements such as samarium and lutetium and rhenium, and these elements, locked within a variety of minerals, also provide atomic clocks that illuminate the deep past.

It is modern Earth magic, but magic of a kind that may be more likely to be possessed by a technologically aware alien civilization than the rougher magic of a history based on the measuring of the shape and distribution of preserved corpses within hardened layers of sand and mud. For most rocky planets, not to mention the abundant debris represented by asteroids and meteorites, are considerably better endowed in atomic clocks than are most Earthly sediments, altered and rotted as the latter are by our Earthly atmosphere. Radiometric dating is something one can imagine as a standard inter-galactic technique, even if our future archivists may be a little perplexed at how difficult it can be to apply to the surface rocks of this planet.

The geological timescale is today being slowly, painstakingly ground out as more and more strata—mostly the volcanic ash layers they contain—are analysed by the isotope-counting machines. It is now all a question of refinement. It has been known since the middle of the twentieth century that the astonishing outbreak of multicellular life-forms at the beginning of the Cambrian Period took place around half a billion years ago. That particular threshold was, until a few decades ago, thought to have been crossed around 600 million years ago. The best recent results now suggest that it took place a little later, around 540 million years ago. And so it is with the rest of the geological column. There have been adjustments, but no great scientific upheavals that might allow, say, the dinosaurs to coexist, Hollywood-style, with humans (or with future alien visitors), or for them to have evolved billions, rather than millions, of years ago. There has simply been a tighter and tighter temporal grip on when the great (and small) events of the geological past took place. It is difficult to over-emphasize this certainty we have of the bones of deep geological time.

So, complex life began a little over half a billion years ago (or a little over five-eighths of a billion years ago from the vantage point of our chosen time in the future). This is pinned in time, quite securely, by a reading of some of the many billions of atomic clocks imprinted in the fabric of the Earth. We assume, by the way, that our investigators will adopt the Earth year—or a single revolution of the Earth around its sun—as a standard local time unit. For Mars, astronomers today use the Martian sol as the local day-equivalent; it is uncomfortably close to an Earth day, being a little under forty minutes longer, while a Martian year is a little over 668 sols. Whatever the time units used, though, half a billion years ago may be considered a little late in the history of a planet to have a major biological revolution. For the entirety of Earth history will almost certainly have been rapidly pegged by our extraterrestrial explorers at a little over four-and-a-half billion years, by radiometrically dating samples of the debris left over from the formation of the solar system: the asteroids and meteorites; this would be the obvious thing for experienced space scientists to do. In this way, the enormous extent of that part of Earth history before multicellular creatures appeared will emerge from the mists of the deepest of planetary time.

On the surface of the Earth itself almost nothing tangible remains of the first half-billion years of the time (evocatively termed the Hadean) when the Earth coalesced from its primordial components, probably by multiple collisions of planetesimals; of the time when, once grown, it underwent an 'iron catastrophe' in which its basic components separated out into an iron-rich core and a silicate mantle; of an intense meteorite bombardment, of which the greatest strike on the Earth, that of a Mars-sized planet named Theia (in Greek mythology the mother of famous children, respectively the Sun, the Moon, and the Dawn), is the most likely mechanism to have flung out the debris from which the Moon formed; and of the cooling and crystallization of a planet-wide magma ocean to form the beginnings of a terrestrial crust. It is unlikely that much will be found older than the few zircon crystals, over four billion years old, that have been recovered as grains washed, a little under two billion years ago, from some very ancient rock (current whereabouts unknown) into the sands that became the sandstones of the Jack Hills of Australia. And it is unlikely that preserved entire strata much older than the 3.8-billion-year-old Isua Supergroup deposits of Greenland will be found.

Nevertheless, those first sedimentary rocks show a recognizable Earth, with liquid water at the surface. And then, or very shortly afterwards, there appeared life—microscopic bacterial life. Odd clues to this life may be seen in the proportions of different kinds of carbon isotopes in rocks of this age, a signal that microbes had been sifting these into different compounds as they metabolized. There are also limestones showing fine, irregular layering that defines football-sized mounds and pillars and columns; these are stromatolites, structures formed by sticky microbial mats that grow incrementally by trapping fine sediment and then overgrowing it, layer by layer (living stromatolites are found even today, rarely, normally colonizing hostile environments—hypersaline ones, say, that exclude the multicellular creatures that would normally eat such microbial structures). And occasionally, preserved within the finest-grained of flint-like silica-rich rocks, the microscopic outlines of single cells or strings of cells.

Earth's living history is almost all of such organisms. It has, in reality, been mostly a continuous, unbroken Slimeworld: three billion years or more of microbes that coated all of the sea floor in sticky mat-like aggregates, that floated in the sea, that almost certainly extended into rivers and lakes, and lived in caves and in the primitive soils of that early landscape. Microbes would have colonized rock fractures and the pores between sediment grains in rocks as deep as a kilometre below the solid surface, as they do today. They would have included species that could live above the boiling point of water, and those that could survive deep freezing, for between them the microbes can turn almost any chemical reaction to their advantage. Microbes are still the dominant life-form on the planet, in terms of their numbers (we each play host to some five billion specimens of *Escherischia coli* within us) and of their metabolic effect on the Earth. Far more so than multicellular creatures, they are vital to maintaining the conditions for a stable, functional global ecosystem.

Our extraterrestrial investigators will likely know this, and so will be looking out for microbial remains within the Earth's oldest rocks—and will likely find them, for microbes and microbial mats are likely to be as universal a form of life as is present on Earth. Many more planets are likely to harbour microbial life than are home to the large multicellular organisms. We are now searching for microbes on Mars, after all (and more likely than not a few will turn up), rather than for the remains of the ancient and malevolent civilization of H. G. Wells's imagination.

The future explorers will also be aware of the effect that microbes can have on any planet upon which they appear, or perhaps upon which they find a home (for some microbes can survive freeze-drying in a vacuum, and ideas of them being carried through space are not necessarily fanciful). When microbes are found on Mars, it will be interesting to see whether they possess DNA; if so, that of course might make us Martians by descent. The early microbes would have made a living by metabolizing the raw materials about them. Microbes are adaptable and evolve quickly and—so to speak—learn new tricks, by virtue of trial and error carried out by many trillions of individuals producing several generations each day; witness how quickly pathogenic microbes have become resistant to antibiotics in recent years.

Then, a microbe learnt a new trick that changed the world. It set in train the first major global ecological catastrophe, that ultimately—at least from an anthropocentric perspective—proved a boon; indeed, it made possible an anthropocentric perspective. It evolved a means to harness the energy of sunlight and to use this to combine two of the most abundant raw materials then present—carbon dioxide and water—in order to manufacture carbohydrates, the building blocks for its own growth. But, the by-product of this reaction is fiercely reactive, and was about as toxic to virtually all organisms then as chlorine is to us now: it was oxygen, and the microbe had invented photosynthesis.

The oxygen was simply excreted into the oceans, oceans that up until then had been anoxic (devoid of oxygen) and therefore possessed a significantly different chemistry to today's oceans. In today's oxygenated oceans any iron is present as, or immediately converted to, the oxidized ferric form Fe^{3+}; this is virtually insoluble and forms a hydroxide—rust, essentially—that settles with fine-grained mud particles on to the sea floor. But in anoxic water, iron becomes the reduced form Fe^{2+}, which is much more soluble in water. Thus, the early Precambrian anoxic oceans were iron-rich, and when oxygen appeared on the scene, the oceans formed the first line of defence, the means of absorbing and neutralizing this poison.

For over a billion years, between roughly three billion and two billion years ago, rust precipitated on to the ocean floor, interbanded with layers of silica, to form the distinctive strata of the Banded Iron Formations, so ubiquitous in Precambrian terrain that they have their own acronym, the BIFs. We now exploit them for almost all the iron and steel we use, for they dwarf any other iron ore deposit on Earth. When our extraterrestrial explorers also need iron for their expanding colonies, it will be to the BIFs that they will turn; and they will no doubt soon form a conclusion as to how these deposits arose, as by-product and witness of the revolution that led to an oxygenated Earth.

As the oxygen spread across the oceans, and slowly began leaking out into the atmosphere, the *ancien régime* microbes, those that could only function in an oxygen-free environment, retreated as a protracted hecatomb of their numbers and of their diversity began to take place. They are still here on Earth, of course, but they are now restricted to those parts of it where oxygen does not reach, within deep rocky crevices and stagnant organic-rich muds. These environments are not rare, admittedly, but the anoxic bugs no longer have the planet to themselves. The microbes evolved, inevitably, forms that not only tolerated oxygen but could exploit this highly reactive stuff as a marvellous energy provider, using it to burn hydrocarbons to obtain energy.

And so there was another billion years of an increasingly oxygenated world, with the microbes adapting to—even while controlling and modifying—this new and changing environment. A stable world, this, of the later Precambrian? Perhaps not. There were swings in ocean chemistry, as seen in isotope ratios preserved in the rocks, more extreme than anything seen subsequently. The celebrated 'Snowball Earth' glaciations took place then. They may have been slushballs rather than snowballs. Nevertheless, they were more extensive than any of the later glaciations, and their terminations may have been dramatic: unusual dune-like structures in the strata that immediately overlie the glacial deposits have been interpreted as the traces of hyper-hurricanes.

Then, 540 million years ago, geology changed. At the beginning of the Cambrian Period, there was an explosive origin and radiation of multicellular animals. It was a striking event, one that seems ever more extraordinary as we study it more closely. It is the most obvious stratal boundary on Earth, now and for the foreseeable few billion years, barring catastrophes so extreme that one may choose to ignore them. It will certainly be a focus for study for our extraterrestrials. They may well be as puzzled as we are.

Consider this: all the major groups of animals on Earth (including, as we now know, the vertebrates) originated within a mere twenty-million-year span. There are no obvious precursors, yet the early Cambrian animals already possessed 'advanced' features. The true metazoans were preceded by enigmatic multi-celled soft-bodied creatures, immobile in life, discovered fifty years ago in Charnwood Forest near Leicester, and almost simultaneously in the Ediacara Hills of Australia. Nevertheless, these were almost certainly failed experiments, and unrelated to animals and plants as we know them.

Was there an environmental trigger for the Cambrian explosion? A rise in oxygen levels has been mooted (for the metazoa are energy-demanding and oxygen-hungry). Or perhaps it was something to do with the development of offensive or defensive armour and the beginning of an arms race that drove metazoan evolution subsequently, and is still driving it. Many of the early creatures are skeletonized, and there was a geochemical 'phosphate event', with an abundance of this substance, around the beginning of the Cambrian. Or was the trigger biological? I am drawn to the idea, that I first heard expressed by Simon Conway Morris, that the intercellular command-and-control mechanisms necessary to organize cells into tissues, and the tissues into an organism, are so dauntingly complex and hard to construct (and so easy to tear apart by the enterprising microbes) that it needed a few billion years of evolutionary trial-and-error and the right environmental conditions (oxygen and so forth)—and luck, presumably—to hit upon them.

LIFE-TIME

Once the metazoans had arrived, by whatever means, the world changed. From our immediate point of view, and as regards the labours of human geologists, and of their unearthly counterparts and successors a hundred million years hence, proper stratigraphy became possible. That is, a means was put in place by which the history of the world may be minutely, scrupulously analysed and recorded, by exploiting the evolving shapes and patterns of complex multicellular life itself. We say goodbye to the long vistas of the Precambrian, where temporal uncertainties may amount to hundreds of millions of years; and greet the world of the post-Precambrian, that is, of the Phanerozoic Eon (within which we still live and that will still persist—hopefully—when the extraterrestrials arrive), where time is measured in numbers of millions of years, or, increasingly, in fractions of millions of years. This is the historical scale provided by past biology. Geologists call it biostratigraphy.

It is a splendid measuring device. In the history of the Earth, the currency is time, and the routine measuring tool used is fossils. We use these both to monitor the state of health of the past life support systems on the Earth, much as a miner uses a canary, and to state when and how fast those life support systems changed, for better or worse. Radiometric dates are wonderful when one has them, and they give a numerical date, in numbers of years. But not all strata contain material suitable for dating; suitable volcanic ash bands are all too often a little thin on the ground: like the No. 9 bus, they are rarely there when you want them.

Individual animal and plant species—or those parts of them resistant enough to be fossilized—have particular, indeed *unique*, recognizable shapes. A species appears, evolving somewhere from some pre-existing species, and thrives for a while, perhaps colonizing wide areas of the globe. Then a few million years later, it meets with an unfortunate accident such as an incoming meteorite; or the environment changes; or that species becomes overtaken in the evolutionary arms race by something bigger, faster, more aggressive. It becomes extinct—and never reappears. Let us say it possessed a hard skeleton—a shell, or bones, or a carapace. And that it lived somewhere where it could be fossilized—the sea, or a lake, or a delta. Then, in death, and for as near eternity as the solar system will allow, it became a chronometer, a time marker for the strata that enclose its countless fossilized representatives.

As with the appearance of metazoan fossils at the Precambrian–Cambrian boundary, such a fossil species defines the time interval in which it lived. Across the world, strata enclosing its remains show that an interval of approximate time equivalence has been found. For two centuries, this has been the primary key for the reconstruction of the past Earth.

Some fossils are better than others. Dinosaur fossils, dramatic as they are, are fairly useless in this regard. While they may effectively and dramatically symbolize the Mesozoic Era, they are usually not much help in identifying strata as Mesozoic in practical terms, for geologists going about their day-to-day business. Being at the top of the food chain, they were comparatively rare as organisms, and because they lived on land their remains were much more often than not scavenged, decayed, recycled into the biosphere. In a 30-odd-year career as a field geologist I do not recall finding a single dinosaur bone (a few bits of Jurassic marine reptiles, yes, but these do not count as representatives of the *land*scape of that time).

Smaller and less dramatic fossils are much more useful, and my personal tally here probably approaches the half-million mark (not counting the microfossils bound up in the rock samples). Though even here one makes a distinction, preferring some of these more mundane fossils over others. A biostratigrapher wishes upon any species the life of the wilder kind of Romantic poet: early brilliant success, a worldwide reach, and then a sudden death. Early success and extensive reach, so the appearance of its remains in strata worldwide mark a level that is as close to globally synchronous as possible. In reality, the appearance of most fossils is *diachronous* to some extent, as it always takes some time for a successful species to spread across the world. In the best cases, there may be a kind of ecological blitzkrieg, such as the spread of rabbits in Australia or the rhododendron in Wales; in the worst, a species may take hundreds of thousands of years to plod around the globe. And an animal or plant can never colonize the *whole* world, as it will encounter environments too hot or cold or too wet or dry or too salty for it.

The sudden death is devoutly to be wished for. There is nothing quite so depressing to a biostratigrapher as a successful fossil that wholly overstays its welcome, and goes on and on. In my field of graptolites (extinct colonial marine plankton, living in the early Palaeozoic Era) there is the species *Rhaphidograptus toernquisti*, which was highly successful for something in excess of five million years—a very coarse time slice by the standards of those fossils. My heart sinks at the very sound of its name. But then there is, say, the beautifully distinctive *Monograptus crispus*, that thrived for perhaps half a million years only. Find that particular fossil and you have a very precise time-fix for the strata that enclose it. It is a kind of a Thomas Chatterton (or Buddy Holly, if you wish) among graptolites.

There are millions of fossil species, each with their own time limits, their origin and their extinction. A rich skein of timelines runs through the Earth's strata from which one can build a history far more detailed and more profound than is possible for the temporal wastelands of the Precambrian. From this one can reconstruct the history of life—and more—on Earth. *Slowly* reconstruct this history. For it is painstaking work, dealing with fossils, and there are no obvious short cuts, regardless of what technology might be developed. Perhaps our investigators will bring with them rock X-ray machines, and so lessen the amount of sheer mechanical work in digging fossils out of their enclosing rock matrix. Perhaps they will have sophisticated computerized shape-recognition systems to assist them. Such things might help. But they will still need patience, and to put in many years of work to exploit the Earth's biological chronometer.

To add to the tortuous business of identifying fossils and separating them into the categories that we call species, there is the task of placing them into order within a timescale, of establishing which types of fossil are associated with which others in any particular stratum, which occur in underlying strata (i.e. are older) and which occur in strata above (are younger). The biostratigraphic zonation that is used to date the Earth' s strata today has been built up by many thousands of palaeontologists and geologists, working over two centuries and more, writing monographs, constructing fossil range-charts, creating the enormous literature database that any aspiring biostratigrapher must get to know and learn to use before they can get to work. As a rule of thumb, it takes about five years' solid work to become familiar with the fossils of just one group (the ammonites, say) within the strata of one geological time period (such as the Jurassic). It will be quite as daunting a task for our extraterrestrial explorers as it is for us. But perhaps they will have the enthusiasm and doggedness that are the *sine qua non* of any human palaeontologist.

THE NAMING OF TIME

The sheer abundance of the Earth's geological record is overwhelming. To describe it, discuss it with other scientists, analyse it, there needs to be a common language. Faced with millennia numerous beyond comprehension, the first reaction of a geologist is to become downright formal. That vast span of time has to be classified, formalized, parcelled up into sensible (though still huge) chunks, and then those chunks need to be split into smaller chunks, and then those smaller chunks have to be split still further.

Only then does deep time become, if not comprehensible, at least manageable. What sort of chunks are we talking about? Let us take, for an example, the Jurassic Period. It is to us perhaps the most charismatic period in Earth history, satisfyingly reptilean, atmospherically scary. But just what *is* the Jurassic, other than the temporal haunt of dinosaurians? It is, technically speaking, an interval of time. It spans the time from just before 200 million years ago to 145 million years ago. This is though, a description, not a definition. When the name 'Jurassic' came into being, in the nineteenth century, the name being derived from the Jura Mountains, no one had any idea of how old the dinosaurs were. They might have been a million years old, or stretch back a thousand million years. Furious speculation there was, yes, but no sensible idea on how to arrive at an age in numbers of years. That only became possible in the early twentieth century, when the discovery of natural radioactivity led to the first development of radiometric clocks. What had been established, the best part of a century beforehand, was that the Earth had had a history, of events great and small, which could be read in sedimentary strata all around the globe. Then, well before anyone knew the length in numbers of years of that history, its major chapters were assembled, essentially in the same order that we use today.

The soul of the Jurassic is these days a little hard to grasp, because we now know far too much about it, with literally millions of fossils and rock specimens catalogued in museums, on which thousands of chemical, isotopic, and mineralogical measurements have been made, enabling the history of this sixty-million-year period (as we now know) to be charted in detail. Already the Jurassic can be split into over a hundred smaller time slices, and with further study we may be able to distinguish thousands of time slices (even so, each of these will be longer than, say, the gulf of time that separates us from the builders of the first Pyramids). So one needs to go back, to the earliest days of geology, when only the crudest, the most obvious aspects of our planet's geological history were appearing.

Approximations of geological eras appeared in the mid-eighteenth century. An influential Italian 'geognost', Giovanni Arduino, recognized, then, that there were hard crystalline rocks, often crumpled and sheared and presumably quite ancient; there were also layer-cakes of more or less hard rock strata, variously tilted; then the softer, flatter-lying upper parts of these; and then scatterings of sands and gravels and clays at the Earth's surface. These he termed Primary, Secondary, Tertiary, and Quaternary rocks, respectively (the last two of these terms still surviving today). That remarkable aristocrat, George-Louis Leclerc, Comte de Buffon, thought on similar lines. He wrote in his *Époques de la Nature* in 1788 that it was necessary to *fouiller les archives du monde, tirer des entrailles de la terre les vieux monuments* so as to reconstruct the *différens ages de la Nature*.

So far so good, but it took painstaking work by the early geologists to hone this poetic vision into a functional Earth history. The breakthrough was made by a civil engineer, a surveyor for roads and railways, with a keen curiosity allied to his professional observations. At the turn of the eighteenth century, William Smith realized that the strata exposed in his road and railway cuttings formed a sensible pattern: and, over much of his life, in an astounding labour of love, traced these over much of England to produce the world's first geological map. In doing this, he devised the discipline of lithostratigraphy—the distinguishing and characterization of rock bodies. He worked out which stratum lay on top of which, drawing cutaway diagrams—geological sections—thus adding both the third dimension and, implicitly, also, the fourth dimension, time, to the flat two-dimensional surface of the map.

William Smith paid attention to detail, and this made him, in practical terms, the world's first biostratigrapher, too. He noticed that the strata contained fossilized shells—sea-urchins, ammonites, and so on—and each stratum appeared to have its own particular fossils, different from those of other strata. This was useful, because as he traced the strata across country, some of the strata changed from one type of rock into another. Nevertheless, some, at least, of the fossils persisted, assuring him that he was dealing with, essentially, the same entity. One road from here led—and is still leading—to ever greater detail, and to ever further subdivision of those fossil-based time slices. This road, though, leads ever farther from any kind of clear and simple lingua franca of time that is accessible to all. It was, conversely, the grander, overarching patterns that were to provide a common language that scientists and Hollywood film-makers alike could employ.

William Smith's scheme was surprisingly sophisticated, with more than twenty-five different groups of rock strata. It was an effective way of classifying the geology of regions. But, it did not provide a sensible framework for the geology of continents, let alone of the world. There were just too many rock units (and it was obvious that many of them could be subdivided, virtually *ad infinitum*). And, being essentially rock-based, once the English Channel was crossed, it became obvious that the rock layers themselves were changing, often beyond sensible recognition. But, the fossils persisted, and looked at broadly, could be grouped into broader units, that spanned several of the individual rock formations. This opened the way to a larger-scale classification of time.

The great system-builders of the day, building on William Smith's conceptual and practical breakthroughs, then began to carve out and give names to the great chasms of geological time, much as their fellow empire-builders were carving up the world between them. One has to say that the geologists went about things in a more sensible way than did the warrior-politicians, who drew their political boundaries straight across mountain and rainforest, across cultural allegiances, with bloody consequences which persist with us today. The geologists, though, looked for patterns, the deepest natural patterns in the history of the Earth that, for most practical purposes, were the patterns of life.

What pattern of life, then, was ascribed to the Jurassic? The dinosaur skeletons, then as now, attracted the excitement, but it was the small fry that acted as signature: the ammonites, those spiralled shells of which the pearly nautilus is the last relative, the cigar-shaped belemnite, the distinctively coiled oyster-shells, and much else. So common that local people gave them names—snakestones, elf-candles, devils' toenails—long before the science of geology developed, they mark out Jurassic strata from Yorkshire to Dorset, in the Volgian Basin of Russia, in the Jura Mountains and the Ardeche, in British Columbia and Nevada; indeed, around the world.

The grand patterns, though, go beyond fossils. The overall appearance of the strata also helped the early geologists recognize the great divisions of geological time. Go to the striking river cliffs beneath Aust Cliff, for instance, under the Severn Bridge. The strata forming the lower parts of the cliff are bright red; those overlying them are cream-coloured; those forming the top of the cliff are deep grey, almost black.

Throughout much of Britain and western Europe, the red strata, such as form the base of the cliff at Aust, are termed the New Red Sandstone. They belong to the periods that preceded the Jurassic: the Permian and Triassic, when there were widespread terrestrial landscapes, with baking deserts and dried-out salt lakes. The cream-coloured strata begin to show a marine influence, a hint of what was to come, and represent fossilized, brackish lagoons. The Jurassic strata that lie over these are compacted black muds, the earthly record of a sea that swept over the land, carrying the newly evolved ammonites with it. For sea level was rising across the world, and drowned many landscapes. The combined signal of sea level rise and new, changed life-forms was a striking signal to the earliest geologists that here was a natural boundary, a turning point in Earth history that could be widely recognizable.

It seems rather an accident that this time period was first defined in the Jura Mountains. Had the English geologists who had created the early framework been a little keener to build their own empire, then they would surely have been the first to identify and name this period of deep geological time. It would surely, then, have been called the Dorsetic, or perhaps the Cotswoldian, or maybe even the Yorkshirious. Would these names have the same poetry or resonance? Would blockbuster films of malevolent saurians have been created, to quite the same worldwide acclaim? It is a fine point of socio-geological speculation.

Words are important to humans, and hence our need to label slices of time: Cambrian, Silurian, Jurassic. Perhaps they will also be important to the future explorers of this planet. Or maybe these unearthly visitors will prefer numbers. For it is likely that they will already be familiar with a different chronometer, one that is quite novel to human geologists, and that promises another revolution in geological timekeeping. This chronometer comes from outer space. It is called the Solar System.

ASTRONOMICAL TIME

This yet more precise chronometer for strata is emerging, almost literally, from the harmony of the spheres: from the shifting interplay of gravity and angular momentum in the Solar System. We have already made its acquaintance. It is the astronomical variations in the Earth's spin and orbit, so patiently worked out by James Croll and Milutin Milankovitch. The discovery of a profound astronomical influence on the Ice Ages and on the ups and downs of sea level has provided a fundamental blueprint for understanding past, present, and future climate, arguably the most pressing question our civilization has to understand and resolve.

Almost as a throwaway, though, it gives us a key to precisely calibrate the history, not just of the Ice Ages, but of the Earth itself. By choreographing the atmosphere, oceans, ice caps, and life itself on a scale grand enough to leave indelible imprints in the Earth's strata, the variations in spin and orbit provide a steady ticking pulse through deep Earth time. The kind of stratal patterns they generate provide barcodes of geological time with which geologists look set to chart the Earth's history with remarkable accuracy for hundreds of millions of years back in time. If these patterns can be traced back reliably and consistently, the use of fossil zones to calibrate Earth time may seem akin to a crude metre rule by comparison with that most delicate of millimetre scales provided by Milankovitch's marvellous mechanism.

This realization in the development of our own science was slow in coming, partly because Ice Age research has long represented a special enclave, well-nigh a ghetto, within Earth history: the timescale is so short, the icehouse environments so extreme, the ecological changes so marked. Scientists can spend a lifetime working within the strata of the Ice Age, with little need to look at older rocks. And for 'classical' geologists, the complexities of Ice Age science have often seemed alien and unfamiliar.

Nevertheless, the worlds of ancient geology and recent geology are now showing greater parallels. The deep sea floor sediments from which the Milankovitch patterns are best derived carry on down and down, far into the realm of deep time. By drilling deep boreholes through the ocean floor one can, here and there, reach layers of oozes that were laid down when the dinosaurs were still roaming the land surface. And one myth, at least, was laid to rest almost immediately, as scientists compared the barcode patterns that were emerging from the sediment layers: that of the incompleteness of the Earth's geological record, with the well-worn cliché of the Earth's strata resembling a book from which most of the pages have been torn out. Not true—of these deep-sea strata, at any rate. The Milankovitch signals plucked from the deep ocean could be matched, well-nigh faultlessly, from one borehole to another, even if they were thousands of kilometres apart. So, there *were* successions of strata stacking up, here and there, to preserve virtually uninterrupted records of long time intervals.

Next, the power of this signal as a geochronometer of geological time of unprecedented accuracy was tested when, still in the geological instant of the Ice Ages, the astronomical signal was compared directly with radiometric dating as to which provided the better timekeeper. The record of climatic changes from the deep-sea cores were simply laid alongside the calculated astronomical rhythms of wobble, tilt, and orbital stretch. They fitted like a glove. But, they didn't exactly fit the estimates of age of the deep-sea muds obtained by radiometric dating. The discrepancy was just a few per cent, but seemed consistent. Which was right? More ocean-floor cores were examined. The astronomical rhythms showed the more impressive consistency, and the orbital chronometer won out. Its ability to resolve fine intervals of time was breathtaking: events at a million years' distance could be correlated with errors down to as little as 5 000 years—a better than 10-fold increase in precision.

So have radiometric dates and fossils become redundant? Not in the slightest, because the Milankovitch barcodes *by themselves* are scientifically indigestible. The pattern of the beats is too monotonous to use in isolation, and it is too easy for a few beats within an individual stratal succession to be skipped (if part of a sedimentary succession that accumulated on the sea floor was swept away by a submarine landslide, for example). But held within a tight framework provided by fossils and decaying atoms, the astronomical barcodes come into their own. The last few years, indeed, have seen a race backwards through the geological column to pluck an unbroken astronomical signal from ever older rocks.

Just how far back can this chronometer go? Here, one comes up against the limits of mathematical calculation. The motions of the planets with their myriad gravitational interactions are, ultimately, chaotic. Beyond 30 million years, it is impossible to reconstruct the basic interactions of orbit and spin. But there is a saving grace that enables a signal to shine through the noise for much longer: resonances which develop within the planetary motions. These flip from one resonant mode to another with long repeat intervals and so provide slower, more stable pulse-beats. In particular, a 400 000-year pulse in the orbital stretch cycle should be recognizable in the rock record, and serve to calibrate the Milankovitch patterns back in time, for 100 million years or more.

There is another problem in reading astronomical signals from so long ago. Thirty million years ago there were ice caps big enough to influence the oxygen isotopic signal of the ocean and provide signals clear enough to be read from the strata. But, farther back, in the greenhouse world of the Jurassic and Cretaceous periods, warm conditions stretched around the globe and there was little or no ice, and hence no obvious way for the Milankovitch signal to express itself.

Yet express itself it does. The Chalk strata of the Cretaceous, some one hundred million years old, often show rhythmic banding. As far back as 1895, a remarkably prescient American geologist, Grove Karl Gilbert, had suggested that each stripe was produced by wobbles of the Earth's spinning axis that somehow, he said, affected climate or sea level to change the rock type that accumulated on the Cretaceous sea floor. Gilbert's ideas fell out of favour for the best part of a century, when the general distrust of Milankovitch-type mechanisms reigned.

Modern researchers such as Andrew Gale, though, examine and interpret the rhythmic pattern in the chalk strata of the Cretaceous in much the same way as Gilbert did a century ago, only this time with enough mathematical rigour to implicate clearly, rather than hypothesize, the Milankovitch patterns. Collecting data from the rocks themselves can be simplicity itself. Take photographs of the rock strata and analyse the relative darkness and lightness of the chalky rocks—in reality a reflection of how much mud each stripe contains. And then, the rather more tricky bit: mathematical analysis of the patterns to see just how closely they approximate to a Milankovitch signal.

In the Chalk, the 20 000-year cycle shines through, together with a 100 000-year signal and, more weakly, 400 000-year and 38 000-year signals. Stratal signatures in different places can be compared. For instance, the detailed patterns of English and Ukrainian Chalk strata, when put side by side, matched so well that an 80 000-year gap in the English rocks—representing an interval of time when no sediment accumulated on the sea floor in this part of the world—could be clearly picked out. Minute changes in planetary motions had left a precise scaffolding of time in these ancient rock strata as well.

These Cretaceous signals—and others from even older rocks—are still isolated segments, 'floating' in ancient time. Where barcode patterns match, they can be used to provide a precise link between one succession of strata and another. But they still cannot be used to find out exactly how old those rocks are, as they are not yet keyed into a continuous astronomical record. Geologists will need to fill in the gaps in the signal to do that. The regular 400 000-year pulse is the most likely candidate to bridge the existing gaps and splice together a record of astronomical time that may ultimately extend back for hundreds of millions of years.

How did Milankovitch mechanisms work in the greenhouse world of the Cretaceous? It was a very different world from ours, with half our present land masses submerged, and the world's deep oceans not freezing cold, but akin to a luke-warm bath. We are far from understanding it. But, stratal patterns have been drilled from the flanks of an early Atlantic Ocean, that was then only hundreds, rather than thousands of kilometres across, and hemmed in by landmasses. In the eastern, nutrient-rich parts of the ocean, organic-walled planktonic algae, the dinoflagellates, thrived. Their decaying remains fell on to a sea floor, using up the oxygen to create stagnant, lifeless conditions. Every 20 000 years, nutrient levels fell, the adjoining continents became drier, and hardier limy microfossils took over the seas. Oxygen returned to the sea floor, and a host of animals—worms, crustaceans, echinoderms—moved in, only to be asphyxiated, thousands of years later, as the next astronomical cycle kicked in. In the central Atlantic, farther from nutrient sources, only the calcareous microfossils lived, and it is variations in the abundance of their skeletons in the sea floor sediments that pick out the astronomical signals. In the western part of this same small palaeo-Atlantic, there is the same pattern, but expressed as alternations of organic and silica-producing plankton, again varying as levels of nutrients, washed in from the land, rose and fell. So: three different parts of a small ocean, three different ecologies and modes of reaction to outside forces—but all three danced to the same astronomical tune.

It appears that, icehouse or greenhouse, we cannot get away from the insistent nudge of our own planetary cycles, pulling our intertwined ecologies this way and that. Even if humanity's best efforts push us out of the Ice Ages, it is likely that what is left of our environment will still be regulated by the criss-cross rhythms of tilt and wobble of the Earth's spinning axis and stretch of the Earth's orbit.

Human scientists have been a little slow in recognizing the astronomical barcodes preserved within the Earth's strata. This is understandable. Our perspective has been that of local observers bound to the surface of a large planet, and so not conditioned to turn an astronomical telescope back on *to* that home planet. We are conditioned to be more aware of influences that lie close to us, such as storms, earthquakes, and volcanoes, than those that emanate from the distant planets.

Our future explorers may catch on rather more quickly. They may well even start consciously and actively looking for Milankovitch patterns as soon as they realize the scale of the stratal riches that encase this planet. For virtually every planet, anywhere in the Universe, will show variations in its orbit around its particular sun, and in its axis of tilt; it would take an exceedingly well-behaved planet not to do so. The Earth, among the planets of the Solar System, is not at all extreme as regards these properties, not least because its axis of spin is more or less anchored in place by its substantial moon, and only varies by a couple of degrees.

Mars, by contrast, has a couple of tiny rocks orbiting it that are called moons only out of politeness, and that are given the impressively classical names of Phobos and Deimos. They are far too small to stabilize Mars's axis of spin that, consequently, varies through time by nearly forty degrees, swinging backwards and forwards between high-tilt and low-tilt states. The resultant systematic changes in sunlight and climate are firmly marked on the only strata on that planet that are still systematically forming: the layers of ice, dust, and frozen carbon dioxide that form the polar ice caps. These, even when examined from satellite, show clear Milankovitch patterns. They will be universal stratigraphic markers, quite literally, far more so than the fossilized presence of hard-shelled dead organisms. The presence of abundant fossils might turn out to be a quite terrestrial oddity.

The Earth's astronomical pacemaker has, finally, a use that is almost philosophical. For, it may give humanity (and perhaps, in the future, the beings of other kinds of civilizations) a proper grasp of the vastness of geological time. For professionals and lay people alike, this immensity has been utterly impossible to grasp, something to be coped with only in the abstract. Years, centuries, millennia even, we can comprehend—but a million years is way off any human scale. Now stack a few 20 000-year wiggles up on a page. A quarter of a wiggle back, the Pharaohs were building their Pyramids. Half a wiggle back, and the ice was retreating, with Stone Age tribes colonizing the newly thawed ground. Twenty wiggles or so back, our species was born. A hundred wiggles ago, the Ice Ages began. Three thousand wiggles ago—you can fit them on a page, if you draw carefully—a meteorite impacted, and the dinosaurs vanished. Here is the key, perhaps, finally, truly, to imagine the past.

So given this playground of petrified life amid the Earth's strata, given the added ultra-high-resolution timekeeper provided by the astronomical stretches and tilts and wobbles, given enough extraterrestrial explorer-scientists working for long enough, given that they have enough enthusiasm (and perhaps enough of a sense of humour): what will they uncover of this last, biologically crowded half-billion-year phase of Earth history?

DYNASTIES AND REVOLUTIONS

It should fairly quickly become clear that life on Earth has followed a dynastic pattern, a pattern discovered by our Victorian and pre-Victorian antecedents. Take first the era started by the Cambrian explosion, where life had an *unfamiliar* aspect. It is the era we call the Palaeozoic. It has an empirical reality independent of classification schemes. It would likely be recapitulated as an entity, a unit of Earth history, from a far future vantage point, just as it forms such a unit today. Its seas were dominated by animals of the same major groups (phyla) that we see today, but in different proportions, and with the main types (the classes) of animals and plants within these phyla differing markedly from today's. They will undoubtedly differ, also, from whatever types of animals and plants will appear on the Earth in the next hundred million years. Among the arthropods, the trilobites dominated for most of the Palaeozoic. Among the shelled animals, it was the brachiopods (lamp shells) that held sway while the bivalves (a.k.a. lamellibranchs, or clams) were still bit players on the sea floors. Corals were abundant, but these were different from today's corals in structure. Life on dry land only really became prominent halfway through the era, and this is the time of forests—not of oak and ash and pine, but of giant clubmosses and horsetails and seedferns. These forests were devoid of flowers, or of birds or squirrels or deer; they were habitat, though, for the first amphibians and reptiles and dragonflies.

The Palaeozoic world was quite unlike that of the succeeding Mesozoic Era. An amateur palaeontologist would have little difficulty telling handfuls of Mesozoic and Palaeozoic fossil shells apart. The rugose and tabulate corals disappeared, and the hexacorals arrived. The ammonites and belemnites swarmed through Mesozoic seas, and may be picked up today in their hundreds on the Yorkshire and Dorset coasts. The brachiopods are represented by the rhynchonellids and terebratulids, now common on any Cotswold hill, rather than by the orthids and spiriferids and productids of the Palaeozoic limestones, but these have now become seriously challenged by the bivalves. Among the very smallest fry, the microplankton developed calcareous skeletons in the Mesozoic, and so created an entire class of rock—the deep-water pelagic limestone—absent from the Palaeozoic.

How, though, does one world pass into another? Slowly, gradually—or abruptly, in a geological revolution, perhaps even in a welter of chaos and destruction? The fossils themselves, augmented by the astronomical pulsebeats, tell the story of themselves and of their changing world at the end of an era. It was a world, in this case, that suddenly became hostile.

The Palaeozoic world fell apart. The way in which it fell apart can be seen wherever there are strata that capture this mass killing, the biggest die-off in Earth history, with some 95 per cent of species disappearing. Volcanism? Cometary impact? Global suffocation? The causes of this extinction event remain unclear, but there are strong signals that the oceans stagnated, and somewhat less convincing signs (such as a tendency for reptiles to become barrel-chested) that atmospheric oxygen was in short supply. Democratically, officially, and formally, by a vote, taken in 2001, this environmental putsch was marked by a 'golden spike' in Meishan in eastern China, a reference level chosen within the succession of strata there, that are held best to define the boundary between the dying years of the Palaeozoic and the dawn of Mesozoic time.

The Meishan section is perched high up a hillside, and, as such, is being eroded. One hundred million years from now, it will be gone. But, equivalents of these strata elsewhere remain deep underground, and some of these equivalents will undoubtedly be at the ground surface, somewhere else on Earth, a hundred million years hence. They will reveal a similar story, though the particular expression of the evidence will be different. No matter. The boundary shows such dramatic changes that this ancient crisis will be remarked upon in any future geological analysis, to work out just how close a planet can get to losing its multicellular organisms, and yet still hang on and recover.

The Mesozoic Era in turn came to an end as suddenly as had the Palaeozoic, though the scale of the calamity was less. There was a complete disappearance here of dinosaurs (well, except for their descendants the birds), of ammonites and belemnites and rudist bivalves. Many other species and genera disappeared. This emptying of ecospace ushered in the world of mammals on land and—as whales and dolphins—in a sea suddenly bereft of marine reptiles; a world in which the brachiopods were now bit players among the bivalves and the gastropods that now crowded the sea floor.

In terms of human-disaster science it is notable for being the first time a major extinction event became founded on solid evidence rather than on hopeful speculation. Prior to the remarkable discovery made by Luis and Walter Alvarez, father and son, the extinctions had been ascribed to anything from sea level change to a sudden outbreak of constipation among the dinosaurs (the ammonites presumably forgoing their normal metabolic practices in sympathy). Their discovery of a 'spike'—a sharply defined enhancement—in amounts of iridium in strata, at the level that was coincident with the extinctions, changed that. Suddenly, hard evidence had appeared—and was bolstered, as the iridium spike was found at exactly the same level in the strata (as far as could be judged) elsewhere around the world. The cause of the anomaly might be disputed: the Alvarez team ascribed it to a massive meteorite impact (with the candidate crater later being found beneath the Yucatan peninsula of Mexico), while others suggested massive volcanic outbursts (now preserved as the voluminous basalts of the Deccan Traps in India) as a source. But here, at last, there was a trail that could be followed, further clues to be uncovered, hypotheses to be exploded or supported on the basis of observations and data.

More, the iridium discovery focused attention on the event itself. Were the extinctions sudden (spanning centuries, or years, or days) or gradual (over some considerable fraction of a million years)? This distinction is surprisingly hard to make from the fossil record, and the reawakened controversy led to exhaustive collecting of fossils from stratal sections around the world that captured evidence of this event. It was a key component in the formal definition of the boundary that marks the end of Cretaceous time, which is placed at a level in strata at El Kef, in Tunisia (again chosen as the world's best example after *much* discussion). The latest assessment of this boundary notes that the El Kef level is best regarded as coincident with the Yucatan meteorite strike; this is a quite remarkable statement for a science in which notions of cause-and-effect, *in particular* with regard to the formal positioning of geological time boundaries, are normally treated with extreme caution.

Our present era, the one ushered in by those tumultuous events, represents a world that has now persisted for some 65 million years, a world in which the Earth's biota adapted to a climate that over time, stepwise, began to chill until massive ice caps grew, first over the South Pole, and later on some of the landmasses close to the North Pole. There have been minor extinctions and evolutionary bursts among living organisms, but nothing to compare with the great extinction events of earlier times—not even during the crazy climate switchbacks and sea level changes of the Quaternary ice ages.

And now, in our present time, the world is changing from year to year, as humanity takes over the Earth's surface and adapts it to its needs. How much of that change is only skin-deep, as far as the Earth is concerned, to disappear in a generation or two? How much will last longer—a millennium, say? And how much of that change is being written into the fabric of this planet, to remain detectable in a million years, in ten million years, or perhaps a billion years hence? To become as near eternal as anything in this solar system can be? The answer will be written in the strata. One will need only to find the message left by the human race, and then to decipher it. The future Rosetta Stone to the human empire will be, at first, a record of how a planet changed through our activities.

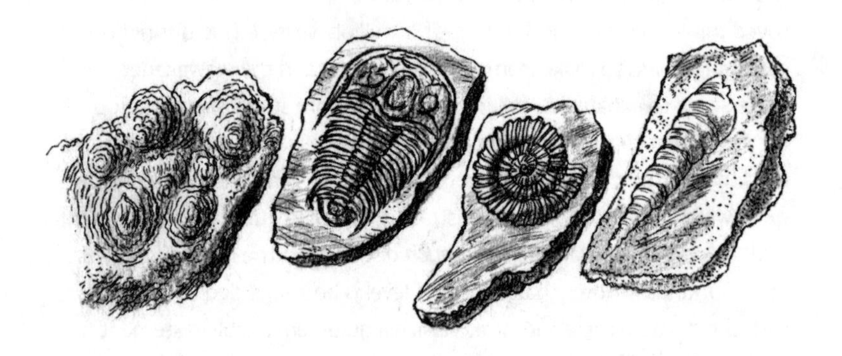

Echoes

It seems this planet has not always been so benevolent to life. We have finally built some sort of a history of these organisms. It is a slow and clumsy business, and our technology has often been of little help. Yet our scientists improvise enthusiastically and tirelessly. Such primitive data-gathering seems to be to their taste. We have argued fiercely over what the data means, but I have now been convinced. There is strong evidence for severe perturbations that drastically reduced the diversity of these life-forms. It is still hard to tell what these perturbations were, but mostly we choose from among the obvious candidates; at least in every case except one.

IN SEARCH OF A HUNDRED MILLION YEARS AGO

It would seem to be like searching for a needle in a haystack. One hundred million years on, strata containing recognizable fossils of multicellular creatures will extend through a 640-million-year span in time, and will have piled up, in total, to many kilometres in thickness. Somewhere in those endless stretches of rock might be the layer in which traces of humanity may be preserved, awaiting discovery by our curious visitors.

Would they happen upon this by chance? Or would they be led to it by following a trail of clues, much as a detective is led to the perpetrator of a crime by the ripple effects of the act itself: the wealth gone astray, the scattered victims, the damaged property, the spree in Monte Carlo. In the case of humanity, there have been victims, and damage, and stolen wealth. It has been a quite singular felony. It will leave echoes, collateral changes, that may act, in the far future, as signposts. Some of these may give clear directions to a heist that was quite out of the ordinary. On a planetary scale, in fact.

But first, how thick a section of strata should we take, as the geological target to be searched for, analysed, interpreted for signs of a vanished civilization? One might start by taking ten thousand years' worth. To us, that represents a gigantic stretch of time (just imagine trying to peer ten thousand years into the future). Geologically, it barely counts. There are a hundred such intervals in a million-year span, and a million-year span represents the small change of geological history. Nonetheless, there are good reasons for choosing this duration, for it represents the span during which human activities can be said to have left a detectable imprint upon the geological record—an imprint beyond the odd vanishingly rare bone of an obscure bipedal hominid. Ten thousand years ago, half of the large mammals of the Earth abruptly disappeared, and it seems increasingly likely that this disappearance was mainly the result of hunting by humans. Now, large mammal bones a hundred million years hence are likely to be just as rare as dinosaur bones are today, and so this disappearance will not be immediately obvious; but, the abrupt cutting-off of several long-established lineages would eventually attract the attention of the vertebrate palaeontologists of the far future. It is humanity's first real footprint upon the wider world.

Cut the span in half to five thousand years, and the effects may be a little more obvious. At this point, many humans had moved on from a life of hunting and gathering, and settled agriculture had begun in earnest. Here and there in the world appeared the patches of cultivated ground, replacing wild woodland and grassland, and growing ever larger, eventually to take over whole countries.

The real global impact, though, struck with the coming of the Industrial Revolution, when both human numbers and human exploitation of materials, energy, and land began to climb steeply, exponentially—as they are still climbing. The span of real human control on the emerging geology is, therefore, not much more than a couple of hundred years. As regards geological time, that is minuscule. In most ancient strata, such an interval is seemingly instantaneous, barely distinguishable from that of a meteorite strike.

So, the human stratal interval, as measured by time, is puny. But where will it be? And physically what kind of thickness might it attain? The best guide to this sort of question is not to look forward a hundred million years from now, but to look back. If we locate the strata that formed between one hundred million years ago and one hundred million and ten thousand years ago, and chase these rock layers across land and sea, we might gain a reasonable idea of what kind of evidence will be available to our future chroniclers.

Geologically, when, exactly, was one hundred million years ago? Well, *exactly*, we do not know, but we can make a reasonable estimate: to within, now, probably a million years or so. The most recent estimate places us very near the beginning of the Cenomanian Age of the Cretaceous Period of the Mesozoic Era. This is within the time-realm of the dinosaurs and large marine reptiles, of the ammonites and belemnites.

Sharpening the focus a little, it is about 35 million years (again, give or take a million years or so) before the sudden extinction of the dinosaurs, that was caused—or hastened—by the calamitous impact of a ten-kilometre-radius meteorite into what we now know as the Yucatan peninsula of Mexico.

In terms of strata, there are few levels that are more convenient. One hundred million years ago represents a level that is very near the base of that thick, highly distinctive, dazzlingly white and pure limestone that we know as the Chalk. Our distant ancestors knew it as the source of the flint with which they made their tools and weapons. For our medieval forebears, the ability to recognize a piece of Chalk was one of the five tests used to tell the sane apart from the insane. In the Second World War, the white Chalk cliffs of the south coast became a symbol of national defence.

Geologically, the Chalk is one of the truly worldwide stratal units. Normally a few hundred metres thick, it extends across Britain, northern France, Germany, Poland, Russia, China, the Americas, Australia. It marked a protracted global event, the greenhouse world of the late Cretaceous, when ice caps were virtually (or absolutely) non-existent, when low-lying continental regions were flooded by a sea in which billions of microscopic planktonic algae—coccoliths—teemed in the sunlit upper layers. Coccoliths possess skeletons of calcium carbonate (that are wonderfully elaborate when viewed with a scanning electron microscope); on death, these skeletons sank to the sea floor to build up the layers of calcareous ooze that, once hardened, were to become the Chalk.

The Chalk seas persisted over Britain for about 35 million years, to the end of the Cretaceous. The Chalk strata are about 200 metres thick, and the sedimentation was more or less continuous (though the uppermost Chalk in the UK was eroded away in early Tertiary times and so, alas, the iridium-rich boundary layer is absent from mainland Britain). Thus 10 000 years' worth of Chalk, on average, amounts to a little over 5 centimetres. And 200 years' worth is about 1 millimetre. That is a microscopically thin target.

Chalk, like most deep-sea oozes, accumulated slowly, and a small thickness encompasses a lot of elapsed time. Other strata can pile up much faster, making the target larger. Where the Mississippi River now empties into the Gulf of Mexico, much of the sand and mud that the river has been carrying is dumped at the coast. This has built up the Mississippi delta—a wedge of sediment reaching 100 metres thick and extending over many thousands of square kilometres—in a little under 10 000 years, since sea level rose at the end of the last glaciation. In time, though, that thickness will diminish to perhaps half of that value as the muds become compacted.

Fossilized deltas comparable to the Mississippi are present throughout the geological record. So the target strata will thicken and thin, depending on what kind of environment they represent, and how much sediment is being supplied to it. In many places strata of a particular time span will be absent altogether, where an area was then being eroded rather than accumulating sediment. Where such gaps in the record occur, no evidence of climate or environment or life (human or otherwise) can be preserved. Most of our land surfaces, for instance, will suffer this fate, as well as areas of sea floor swept clean by strong currents.

Still, given the target stratum, thick or thin, just where will it be? Again, let us take our basal Cenomanian strata, that rest upon strata laid down during the previous Age, the Albian, that include strata such as the Upper Greensand and the Gault Clay. Where are they now?

A geological map will show the surface outcrop of the Chalk. In Britain, it is typically shown as a green stripe that winds from the Yorkshire coast, where it forms cliffs from Flamborough Head to Bridlington. It disappears under the waters of the Wash, reappears at Hunstanton, travels down East Anglia past Newmarket and then Luton, and on via the Chiltern Hills to Salisbury Plain. From there, one branch goes south to Lulworth Cove, while two more track eastwards to form the North and South Downs before their outcrops disappear into the English Channel.

This stripe on the map marks where the Chalk, uplifted and tilted (mostly by just a few degrees) by ripples of the Earth movements that raised the Alps, has been cut across by erosion to lie at the present land surface.

Over most of eastern England, the western margin of the green stripe marks the base of the unit, where it lies upon older rocks. Farther west and north, there are only the older rocks—those of the earlier Cretaceous and then, ever farther west, those of the Jurassic, the Triassic and Permian, the Carboniferous, across to the older Palaeozoic and Precambrian rocks of Wales and Scotland. Across this western terrain, the Chalk once extended widely. But it has gone forever, stripped away by the wind and the rain of many millions of years of erosion. Here there is no chance of finding our 1-millimetre-thick hundred-million-year-old layer near its base, nor will there ever be. And this area of loss is increasing, infinitesimally slowly, year by year, as erosion proceeds.

To the east, our thin layer is first buried under more layers of Chalk, and then, when the green stripe gives way to stripes of other colours on the map, it lies under younger rocks as well, the barely hardened sands and muds of the Tertiary Period. In central London, it is the London Clay (appropriately enough) that lies at the surface, with our hundred-million-year layer lying some quarter of a kilometre deep, far below the London Underground, reachable only by deep boreholes.

Farther east, under the North Sea it descends deeper still—to some two kilometres below the surface, for there the crust has subsided strongly, and even the last couple of million years are represented by half a kilometre's thickness of strata. And then it ascends, courtesy of the tectonic escalator, to reach the surface in France, and from there to traverse Europe.

And so it goes on, this layer: partly buried underground, deeply or shallowly; partly above the present-day landscape as a kind of ghost layer, to be placed by a geologist's imagination as a plane in three-dimensional space from where it has been permanently erased. Connecting the underground realm and the realm, up in the air, of the ghost strata, is the surface outcrop, the single plane where our selected rock layer intersects the modern land surface. It is here at the outcrop that the stratum can be seen, measured, hammered, sampled for fossils or for minerals.

In places the stratum will be associated with ash layers and lavas, where Cenomanian volcanoes were present. Elsewhere, that stratum will have been caught up in the tectonic equivalent of a train smash, where it became trapped between two converging continents. There were some late Cretaceous sediments of the west coast of North America that suffered this fate, buried to 30 kilometres depth, then exhumed, transformed into the Swakane Gneiss of the Cascades Mountains.

Mostly, though, the rock remains emphatically sedimentary, usually as the typical deep-sea chalk. Where the sea floor was shallower and nearer land, though, it may have been covered in sand rather than calcareous ooze, and so marine sandstones will be found. And then, if one keeps tracking the Cenomanian strata across country, one may eventually encounter the edge of that ancient sea, and find fossilized deltas, river plains, beaches. There are a few early Cenomanian deltas around the world. There is one in Alaska, and another in Alberta, and yet another in Texas (where fossilized soils have been discovered). One in Wyoming was nicely christened the Belle Fourche, of the Frontier Formation, while the Santillana delta of the Cantabrian Mountains of Spain is even more prettily named. There is a Cenomanian delta in South Australia, but it lies offshore and underwater; and a possible example in Poland, that lies underground, beneath the Carpathians.

So it is likely that comparable human-epoch deltas will emerge on the surface. How distinctive will it be, this tectonically warped patchwork planet-wide stratal layer marking human existence? If it will be much the same as the thousands and millions of such layers that will lie above and below it, then it will not attract specific attention from our future explorers. And, in that case, the discovery of the ruins of the human empire could only be found by chance, by the lucky hit of a future geologist's hammer amid the endless strata. In such a case, the chances are that the Earth's future colonizers would remain forever oblivious of their predecessors as a manipulative, colonizing, intelligent civilization—oblivious of *us*, that is.

But the human empire may not represent a brief event, following which life on Earth, after our demise, goes back to normal. The human legacy will likely not be like the ruins of Angkor Wat: abandoned, reclaimed by the jungle and, in a few thousand years, totally absorbed back into nature. It represents a threshold, the transition between the world before humans and that to come afterwards. For our impact has been so great that we have already made Earth history. The world will, quite literally, never be the same again. The question is how that revolution will be recorded.

THE BIOLOGICAL SIGNAL

In the geological here and now, a wave of biological extinctions is taking place. Currently, it is accelerating, quite regardless of any concerns expressed about this. The rainforests, a rallying call for a generation of environmental activists, are inexorably being cut to ribbons. In less than a century most will be gone. Millions of species—lemurs, parrots, beetles, frogs, spiders, orchids, jaguars—are set to vanish. Many thousands have already gone, most disappearing before human scientists can find them, describe them, give them a name.

But will this disappearance leave a signal into the future? On land, there is little means to preserve their remains, and only a few recognizable carcasses are swept into the preserving bottle of the sea floor. Generations of these animals and plants have lived and died and, on death, were simply recycled back into the forest. Once that forest has gone, slashed, burnt, briefly ploughed, then grazed, then abandoned, the earthly remains of these species are reduced back to their component atoms and scattered to the winds. Mother Nature has been bountiful, but Mother Earth's memory may be a little short. So will the evidence disappear, eternally, from the scene of this particular crime?

Not necessarily. Vanished forests can leave time capsules behind—if one knows where to look and what to look for. Not so much, though, in the form of entire fossilized trees or the preserved bones of exotic lemurs. True, layers of coal do represent the fossilized forests of the past, but such remains are generally too crushed, intermingled, and carbonized to allow easy reconstruction. Perhaps as children, you searched in blocks of coal for leaf-imprints and fossil bones. More than likely, you would have been disappointed, as recognizable fossils proved frustratingly elusive. Maybe you learned not to examine the coal itself, but went instead to the spoil heaps around the coal mines, to search through the pieces of shale and sandstone that were thrown out there. On lucky days, rock slabs showing imprints of ancient leaves and fern-fronds might turn up. You would have to search long and hard, though, to find enough fossilized fragments to construct a representative picture of that forest.

The clearest record of the vanished coal forests lay, though, in your hand—but you would have needed more than a hand lens to see it. It is almost omnipresent in the shale layers that are sandwiched between the coal seams. To the naked eye, this evidence is invisible. To render it visible, you need take a piece of that shale, and subject it—*very* carefully, in a controlled laboratory—to the corrosive action of hydrofluoric acid. This is fearsome stuff, which does in practice what home-distilled hooch does in legend: it dissolves away the rock, until all that is left is a dusting of fine powder. Spread a little of this powder on a glass slide and put it under a powerful microscope. Look down the eyepiece, and you will see hundreds of tiny ornate spheres, like beautifully designed Fabergé eggs, but just a few hundredths of a millimetre across. These are fossilized pollen grains, made from some of the toughest, most acid-resistant organic compounds in nature.

Pollen is produced by plants in huge amounts, every season, as all hay fever sufferers are too acutely aware. Each type of tree, and shrub, and grass has its own particular type of pollen. Blown through the air in their billions, only a few of the pollen grains find their target in the stamens of flowers. Most of the rest fall to the ground. Vast amounts are then washed into rivers, and carried out to lakes or to the sea, to be entombed into the layers of sediment; each gram of sediment, and the solid rock that it subsequently becomes, can include many thousands of individual pollen grains. By extracting such fossilized pollen, it is possible to reconstruct the nature of prehistoric vegetation; and, by divining the nature of the vegetation, one can work back to the composition of the plant communities, and to the temperature, rainfall, and seasonality patterns which those plants were adapted to: that is, to reconstruct the course of past ecology and past climate.

It is a skilled and time-intensive business. First, one needs to be able to recognize and distinguish the different types of pollen grain. Then, enough grains—several hundred at least—must be identified and counted in each sample to ensure reasonable representation of the different pollen types. There is a little further adjustment to do, as it is not possible simply to translate numbers of pollen into numbers of plants. Some plants produce more pollen than others, and some types of pollen travel better than others. Pine pollen, for instance, is notorious for its abundance and—courtesy of a pair of 'wings' which gives each pollen grain a passing resemblance to a Mickey Mouse head—its ability to be carried far and wide by wind and water currents. The unwary pollen analyst might reconstruct a whole pine forest from the pollen produced by a few long-departed lonesome pines.

And when all that is done, the process has to be repeated, many times, on samples through whole successions of strata. For while one pollen sample can reconstruct a forest at one instant in geological time, like a still image from a film, it takes many pollen samples to show exactly how that forest changed through time.

Successions of fossilized pollen, extracted from the sediments left by now-vanished lakes and rivers, have been used to chart the course of the Ice Ages. In well-studied regions, such as central and northern Europe and

North America, the many twists and turns of climate have been picked out to show how first tundra grasses, then birch forest, then pines, then oaks and other warmth-loving trees spread to take over land recently vacated by ice, only to be pushed back again as the ice advanced once more. The forests of successive warm phases of the Ice Ages resemble each other in general, but differ in the details of their individual evolution. In one warm phase, hornbeam may be richly represented early in the warming; in another, elm might vie with oak for supremacy during conditions of peak warmth. With care and patience, the different abundances of these tree types can be charted through time, to produce a kind of flickering movie depicting the advance and recession of forests in an ever-changing landscape.

The history of the tropical rainforests is more difficult to establish. These are huge, mostly still unexplored areas, and have yet to be adequately studied as regards the plants and animals that live there today, let alone their fossilized ancestors. Simply analysing the pollen (either modern or ancient) is challenging, not least because of the immense diversity of rainforest plants (which means many more types of pollen for the researcher to learn to recognize) and also because of their tendency to be pollinated by insects, rather than by the wind, which means less pollen from each individual plant. Nevertheless, it is clear that these forests also responded to the Ice Age climate changes: not disappearing completely, but being replaced by grassland and shrinking back to isolated patches when the climate became a little cooler and markedly drier, as the ice sheets advanced from the polar regions.

Here and there in Europe, and around the Mediterranean region, new patterns began to appear in the pollen record from some ten thousand years ago, not long after the last retreat of the ice: the disappearance of forest communities, even though the climate stayed warm, and their replacement by grassland; evidence of disturbed ground seen as the spread of weeds—nettles, thistles, plantain. There is the first appearance of the pollen of undoubtedly cultivated plants—the cereals—too, though this is often muted, as cereals typically self-pollinate, and so do not need to release their pollen over long distances.

It is the unmistakable sign—to us—that the growing human populations were beginning to settle and to abandon their former hunting–gathering lifestyle. And with that came the first major human impact on the landscape, the progressive destruction of the great European forests. Europe is now utterly different from its primeval state, and that difference is clearly seen in the record of pollen preserved in layers of mud at the bottom of ponds and lakes, and also in the muds which have accumulated on adjacent sea floors, for pollen can spread far and wide. And that process, accomplished over a few thousand years in Europe, and over the last couple of centuries in North America, is now spreading over Africa, South America, Asia, as the forests are cut back and replaced by pasture and arable land.

An unmistakable marker for the future? Well, perhaps. The world's forests have been subject, over the past few million years in particular, to the vicissitudes of the extreme climate swings of the Ice Ages. As climate became warmer and wetter in the interglacial episodes, so the forests of the world, by and large, expanded. As climate became cooler and drier during glacial phases, they shrank back, and grasslands expanded in their place. So why would this replacement of one sort of pollen by another in the world's strata not simply be the effect of another natural climatic fluctuation?

Here, our future palaeontologists (who, we assume, will have developed techniques to extract and examine fossil pollen from rocks, just as we have) might notice that this particular flood of grassland pollen would not follow the same pattern as in previous vegetation changes of the Ice Ages. They might see, for instance, that the worldwide deforestation in this case took place not as the climate cooled, but rather as it warmed.

There would be something else strange about this flood of pollen. A sameness about it all over the world, and also a certain poverty. For natural grasslands harbour a diversity of grasses, which is reflected in the natural variety of grass pollen they produce. The pollens from the South American pampas are different from those of the great African savannahs, which in turn are different from those of the Mongolian steppes. But, the harnessing of natural grasses and wild rice to the task of feeding a few hundred million, then one billion, then five billion, then (very soon afterwards) ten billion mouths, has meant that only the most productive strains have been selected and bred and exported, massively, all over the world.

It has been called the McDonaldization of life, and the sudden appearance of floods of identikit pollen of crop plants around large parts of the globe is unlikely to pass unnoticed by our future observers. The signal will be complicated. As wheat mainly self-pollinates, comparatively little pollen is released to the wind and the soil. Maize pollen, on the other hand, is produced abundantly (each set of anthers may produce some 25 million grains) and, as maize mainly cross-pollinates, it is carried far by the wind. Oil seed rape pollen is cursed by hay fever sufferers, but it also is a poor traveller by wind—though much is transported by insects. Rice pollen is thin-walled, so less resistant to decay. The pollen record will be a mirror of our agriculture, but a distorting one.

Other, more complex signals will show through, as minor grace notes: the efforts of horticulturalists, travelling the world, and bringing back beautiful and exotic plants for commercial forests, parks, and suburban gardens. Britain, for example, in pre-human times had some thirty-five tree species, and most of the landscape was forest, from coastline to coastline. Now, only a tiny fraction of the original forests are left. But, these remnants now compete with many hundreds of species of imported tree, a horticultural invasion without precedent: a Harrodization rather then McDonaldization in this case, perhaps. Many of these immigrants only survive, in gardens and greenhouses, through the cosseting of the dedicated gardener. Others vigorously out-compete the native trees. Rhododendron from the Himalayas in Wales; and, in plantations everywhere, the Douglas Fir and Sitka spruce, imported from North America.

Consider also the animal species that have most successfully accompanied humans on their global voyages. Brown rats, that, all around the world, jumped ship on to new landscapes which they mostly found to their liking, full of the egg-rich nests of exotic birds. The birds quickly declined, while the rats, infinitely adaptable, thrived pretty well everywhere. With the rats came cockroaches, and rabbits, and, more consciously exported, sheep, pigs, goats, and cattle. Invertebrates, too: the edible snail, for instance, is not native to Britain. There are some surprises: the parrot flocks of southern

England, and the pythons that have taken to the Florida Everglades like ducks to water. Marine organisms, too, have hitched a ride: barnacles, molluscs, worms, seaweed, plankton, attached to ships or as ballast in them, and transported to new shores and new seas.

The transfer of species globally has become a merry-go-round of living organisms without precedent in the Earth's four-and-a-half-billion-year history. It is this particular, distinctive feature that will make the current mass extinction event different from all those that have preceded it: for instance, that which, 65 million years ago, killed off the dinosaurs and much else besides; and that other catastrophe which, almost exactly a quarter of a billion years ago, killed off an estimated 95 per cent of all the world's species. It is too early to say just how our very own extinction event will compare in magnitude with those preceding it: but, at current rates, another century or two should see us competing with the great extinction events of the past.

The most severe effects seem likely to be on land, where other species compete directly with us for the same living space, and in lakes and rivers. The picture is harder for us to discern in the huge, effectively opaque water masses of the oceans. These are much more difficult to monitor. There has been a widespread impression, until recently, of the sea as a virtually inexhaustible resource, which will forever renew itself (with a few highly visible exceptions at the very summit of the food web such as the whales), no matter how many fishing vessels sweep their way across them. That picture is changing quickly. The whales are not the only top predators to suffer population crashes: populations of most of the major shark species, monitored by painstaking analyses of fisheries data, have plummeted in the past decade, their populations being cut to half, or less. A little lower down the food chain, the plight of the cod populations, which have virtually disappeared off Newfoundland, have become headline news. A decade after all fishing was finished, cod stocks have shown little signs of recovery, which gives a telling insight into the ability of fish populations to bounce back after a crisis. The North Sea cod stocks currently seem to be poised on a similar threshold of sustainability.

For palaeontologists, fish are a little too high in the food chain to be commonly fossilized. If the fish are removed, what happens to the organisms lower down that food chain?—that is, to those that are more abundant and hence more liable to turn up as a fossil in the average rock sample. It is less easy to say how these will fare. However, in the wake of the collapse of the cod (and other commercially exploitable fish species) in the northwest Atlantic Ocean, there have been changes to the food web that seem to amount to a thoroughgoing reorganization. Oceanographers call this type of effect a trophic cascade. In this case, the cascading effect meant that as the big fish were removed wholesale for the table, their place in the food web was seized by their former prey: small ocean-going fish, shrimps, and crabs (seal populations have also boomed, because such organisms are their main food source). These interlopers have since kept their place and the cod have not returned. The change might be effectively irreversible, at least on any timescale that might be useful to us.

Would such changes be detectable? Perhaps they might, because these studies indicate the extent to which humans have refashioned marine, as well as terrestrial ecosystems. The changes may well soon encompass such commonly fossilized creatures as the bivalved molluscs and gastropods, sea urchins, limpets, and barnacles of seashore and shallow sea floors. There are also the potential microfossils—planktonic single-celled organisms such as foraminifera and diatoms, with preservable skeletons of lime or silica, which sink to the sea floor and accumulate in their billions in the bottom sediments after the organisms themselves have died. These are the fossils that really characterize strata. It now seems likely that some will show clearly the impact of human activities, while others will provide barometers of wider oceanographic changes in an undersea world that is warming, and ever more intensively exploited.

So, a century on from now, say, let us take stock of the likely situation. Somewhere between a quarter and a half of the world's species extirpated, most disappearing even before they had been catalogued and given a scientific name. This would include most bird species; most of the larger land mammals and quite a lot of the smaller ones; pretty well all the primates, other than *Homo sapiens*; a substantial proportion of the oceans' top predators; and, probably, a significant proportion of organisms lower down the food chain.

Geologically much will be invisible. The lives, deaths, extinctions, and immigrations of the uplands will leave as little trace as the highland dinosaurs of the Jurassic have left—that is, virtually none. It is the coastal plains, deltas and river basins, and the seas and lagoons that border them, that will carry this message.

How long would recovery take? There are good precedents here. The recoveries of the Earth's biosphere after the major extinctions of the past are now reasonably well understood, at least in broad outlines. Previous recoveries have taken of the order of between one million and five million years, until something like the previous biodiversity has been achieved. This provides some comfort to the ecologically minded, if not on any human timescale. 'Recovery', of course, here refers to quantity of taxa, and not their particular quality. For, all the slow dawns following past ecological meltdowns have revealed new worlds, and ecosystems transformed: the mammals replacing the dinosaurs after the end-Cretaceous extinction, for example. So, the most visible part of our likely legacy is not what we are doing *now*. It is what will arise from the ashes. Will the next phoenix be a turkey or an eagle? We have no way of knowing. One could bet, though, on some of those countless rat populations now scattered on islands worldwide producing some quite extraordinary descendants.

As regards our interstellar explorers, learning on the hoof (literally, perhaps?) the rudiments of terrestrial stratigraphy and palaeontology, the great thing about an ecological revolution preserved within strata is that it provides clear signposts towards the event itself. Within the great stratal successions, one is either below the event or above it: the fossils can tell you that, often almost at a glance. It is not hard, say, even for a school geology student, to recognize an ammonite and say that they have found Mesozoic rocks, or (a bit trickier, this) a nummulite and say that they are within the Tertiary. So, thus guided, one can home in on the extinction event itself, to search for evidence as to just what happened and why. This is what palaeontologists do today, as they seek to unravel the mysteries of the mass dyings of the end-Permian and end-Cretaceous times. It would be a sensible course for our future chroniclers to take. Before that, though, they might see some even clearer signals in the strata that something untoward had taken place on this remarkable planet.

THE REEF CRISIS

Today, the rainforests have a rival for environmentalists' affections that is geologically more durable: the coral reefs. The reefs are now in trouble. If they disappear, their demise will leave an indelible and stark record in the Earth's strata, perhaps the most striking of all our signals to the future.

Like the rainforests, coral reefs are hotspots of biological diversity and productivity and, at first—and second—glance, a bewildering paradox. For they occupy less than one per cent of the ocean surface, yet harbour a large proportion of marine species. They teem with life, yet only grow in nutrient-poor seas. And virtually all the visible life-forms are animals—corals, fish, sea-slugs, clams. So what, and where, is the base of this mighty ecological pyramid?

Two keys unlock the paradox. The first is an enviable efficiency in the recycling of essential raw materials such as nitrogen and phosphorus, and here the parallels with a rainforest, growing on a poor tropical soil, are obvious; another parallel might be with a prosperous economy, with money forever in circulation. The second key is provided by a growing realization of the extent of the hidden plant life: in part as symbiotic single-celled algae, inhabiting the tissues of the coral animals and providing nutrition in return for shelter, and in part as algal strands no sooner grown than consumed by myriad snails, clams, and fish. These algae are photosynthetic, and need sunlight; hence, the most productive part of a reef needs to keep within the sunlit surface waters of the sea.

The reefs are marvellous ecological paradises, but their geological importance is that they are mighty producers of rock in the form of lime—or, more simply, limestone. This lime—chemically calcium carbonate—forms the skeletons of corals, clams, crustaceans, and algae, that utilize calcium and carbonate ions dissolved in seawater. The lime is produced in gargantuan quantities, and can grow upwards by centimetres in a year. The reefs are the fastest growing of all limestone environments, important in themselves as rock producers, but yet more important in that they define the form of those more general—and much more widespread—limestone structures known as carbonate platforms.

Just what is a carbonate platform? A small example is the Bahamas platform, a wonderful place, while it still thrives, to study geological phenomena. One uses the word 'small' in only a relative sense. The Bahamas Islands are small, as islands go: a few tens of miles long, a few miles across, and only a few metres above sea level. But their land surface is only the merest tip of an enormous mass of limestone. Swim or snorkel away from a typical Bahamian island, and for a few tens of kilometres you can be in water not much more than ten metres deep—indeed you can often wade out for a kilometre or two, and still keep your head above water. At the bottom of the water, the sea floor is everywhere of calcium carbonate. You are swimming—or wading—above the platform. Swim a bit farther, and suddenly you are over the edge: the sea floor drops away steeply to between four and five kilometres depth. The whole platform is a monstrous flat-topped lump of lime over five kilometres thick, sitting on a deep ocean floor. Well over a hundred million years old, it has been steadily growing, upwards and outwards, over that time, since the Jurassic Period, when the Atlantic Ocean began to open, commencing the divorce of Europe from America.

The sea floor was in those Jurassic times near sea level, allowing the lime-secreting organisms to take a hold, and grow towards the bright sun that nourished them. And generation after generation of corals, algae, clams, sponges grew ever upwards towards the light, growing on the skeletons of their ancestors as the Atlantic sea floor sank ever lower. And for over a hundred million years—and over five vertical kilometres—they kept pace, to produce the gigantic scaffolding of skeleton on countless skeleton that is the Bahamas platform.

The organisms that built these reefs and platforms are tenacious. They have to be, for they have had to survive everything that a capricious environment could throw at them for a hundred million years: to keep growing, to keep building up towards the life-giving sunlight, as their foundations slowly but inexorably sank ever lower into abyssal depths. In that span they showed an almost limitless capacity for absorbing environmental shocks that threatened to overwhelm them time and again.

Sixty-five million years ago, they must have been decimated by the hammer-blow effect, and the physical and chemical aftershocks, of the 10-kilometre-diameter meteorite that plunged into the Yucatan peninsula on Mexico, which ended—or helped to end—the reign of the dinosaurs on land and of the elegantly spiral-shaped ammonites in the sea. Reefs around the world perished, but the Bahamas carbonate factory recovered in time to prevent the oblivion that waited for it in deep water. Fifty-five million years ago, vast quantities of methane—it is thought—bubbled out of the sea floor, giving a brief but savage twist to the Earth's greenhouse effect, and raising temperatures globally by up to 5 degrees. Many organisms perished, but the Bahamas pulled through that eco-catastrophe, too. More recently, they had to absorb the roller-coaster ride of the Quaternary Ice Ages: one geological moment being stranded high and dry as the ice sheets expanded and sucked water out of the ocean, with life just clinging to the edge of what had become an exposed limy landmass; the next—and more potentially lethal—moment, being plunged into deep and dark water, as the ice sheets melted in a series of catastrophic collapses, each one raising sea level by some metres in a century or less, testing to the full the organisms' capacity for rapid upward growth, to keep within the shallow sunlit waters.

But the resilience of the carbonate organisms is not limitless. The Earth's geological record shows many examples of individual reefs and entire carbonate platforms, some much greater in scale than the Bahamas, suddenly killed off. Some smothered by mud, some poisoned, some heated or chilled beyond their endurance, and some simply overfed by an excess of nutrients, then all finished off by drowning, carried into deep water before any surviving organisms could respond and switch on, yet again, the engines of the carbonate factory. The ocean floors then supported many tropical islands that, like the Bahamas, were vast, ancient, and thriving mountains of lime. But soundings of the oceans have also revealed, in depths of a kilometre or more of water, flat-topped submarine mountains called guyots. These submerged plateaux were once coral islands and carbonate platforms, but they drowned many millions of years ago; overcome by some environmental vicissitude or other, they were carried into sunless waters, into an eternal sleep.

The great extinctions, in particular, were unkind to those mighty if physiologically delicate rock-builders, the reef-forming organisms. Reefs vanished during the greatest extinction of all, at the end of the Permian Period, a quarter of a billion years ago. They fared poorly, too, at the end of the Cretaceous, when it was not just the reef-builders that suffered. The most visible trace of that sudden catastrophe was the abrupt disappearance of most lime-secreting planktonic algae: the settling to the sea floor of their minute limy skeletons, the stuff of chalk rock strata, being halted, over much of the world.

The reefs of the Bahamas, and many others around the world, have withstood many different types of adversity. Most of them, though, are visibly suffering in the face of the multiple onslaughts generated by the human race. Some of the attacks are direct, such as simple over-harvesting of the reef animals, particularly the fish. These do not generate carbonate rock themselves, but they help maintain the environment in which the reef-builders can flourish, and their decimation impacts upon the reefs' productivity as a whole. The heatwaves beginning to be generated in our new, fossil-fuelled greenhouse world are making the tropical oceans uncomfortably warm for the photosynthetic algae which live symbiotically in the tissues of the coral animals. These have taken to abandoning their hosts, making their tissues turn from a healthy brown to a pallid white. The corals, so bleached, lose the main engine of their growth, and dwindle and die. The influx in nutrients from agricultural fertilizers, and mud released as a result of development and the deforestation of the adjacent land, further add to the corals' plight. The nutrients cause massive growths of soft algae—seaweed—that out-compete the corals. The mud is not much bother to the burgeoning seaweed, but can suffocate the finicky corals.

Then there is the dynamiting for fish, the accidental damage wrought by a million scuba-diving tourists, the effects of thousands of oil and chemical spills. And there is one more factor to endanger the reefs: the gradual acidification of the oceans resulting from the extra carbon dioxide released into the atmosphere by human activities. Gradual on human timescales, it is geologically sudden, almost a hammer-blow for the coral organisms.

The writing is already on the wall. Extinction rates among the reef dwellers are on the increase, and whole areas of reef are beginning to die. The coup de grâce may be delivered, as so often in the past, by a sudden rise in sea level, sometime in the twenty-first and twenty-second centuries of the human calendar, as the greenhouse effect begins in earnest, and huge volumes of meltwater pour from the collapsing ice sheets. Healthy reefs could perhaps have coped, as they had done during the sea level changes of the Ice Ages just a few tens of thousands of years back. But the enfeebled carbonate factories of late human civilization may be in no shape to keep up with the rising waters. Most of them may simply drown.

Such an extinction, unlike that of millions of organisms in the rainforests, will not be largely invisible and unrecorded. It will produce the clearest possible message in the Earth's strata: the simultaneous termination, worldwide, the growth of massive white mountains of limy strata, and their replacement, and quite literally their burial, by layers of silicate mud and sand. A student geologist in the first week of term could see it. A student geologist keen on palaeontology would not take long to record the disappearance of myriad marine species that accompanied the striking change in rock type. One who looked a little farther would see that rock strata, deposited elsewhere in the world at that time, bore the unmistakable traces of wider extinctions, on land as well as in the sea. The extraterrestrial explorers of the far future will be intrigued by the coincidence. That will not be the end, though, of the evidence.

ACID OCEANS AND THE LIMESTONE GAP

In today's world, samples of atmospheric air have been trapped, for almost a million years into the past, in the compressed snowfall layers preserved at the base of the Antarctica ice cap (Greenland's ice is not much older, anywhere, than a hundred and thirty thousand years). These trapped air-bubbles date back to just a little after the time of the original snowfall; the air is not finally sealed in until the interconnected pores in the fluffy surface snow are squeezed by the weight of subsequent snowfalls, to form isolated bubbles at some metres depth in the snowpack. The bubbles can be extracted from drill-cores of ice and analysed.

For times before that, reconstruction of the atmosphere is difficult. Unmodified air is not preserved in rock, and so one has to look for indirect effects on the strata. The earliest atmosphere of Precambrian times contained little or no oxygen. This can be deduced from some of the most ancient fossilized river deposits, over three billion years old, which contain preserved grains of certain minerals, such as pyrite and uraninite, that would not have survived exposure to an oxidizing surface environment. But as to the exact composition of that early atmosphere there is still debate regarding, say, the relative contributions of nitrogen, methane, carbon dioxide, and sulphur dioxide.

The appearance of oxygen some three billion years ago was marked by the appearance of rust: oxidized iron compounds first appeared in the strata laid down in the sea and, a billion or so years later, oxygen escaped into the atmosphere in sufficient quantities to rust iron compounds on the land surface, to produce distinctive 'red-bed' strata.

Much later—just over half a billion years ago—multicellular organisms appeared, and this evolutionary breakthrough has itself been linked, a little speculatively, with an atmosphere attaining a sufficient 'threshold' level of oxygen to allow metazoan animals to exist. Once animals and plants invaded and progressively colonized the land, between four hundred million and three hundred million years ago, they triggered further changes in the atmosphere. The Carboniferous forests grew and were buried. Carbon, extracted from the air as the forest trees grew, was locked away underground as coal seams. Thus carbon dioxide was extracted from the atmosphere, and oxygen put back in its place. With more oxygen, the forests became more flammable, and fossilized charcoal is common in the coal seams. With less carbon dioxide, a reduced greenhouse effect made global temperatures drop, and an ice cap grew over the South Pole, then occupied by a conjoined South America and Africa.

These are broad changes, over timescales of tens of millions to hundreds of millions of years. They are still glimpsed rather indirectly, though they involved changes in atmospheric composition vastly greater than any taking place today. So what are the chances for today's changes to be discerned?

On the face of it, this is quite a challenge. Humanity has not altered gross atmospheric composition. Industrial activities have led to local increases in some trace gases: sulphur dioxide, nitrogen dioxide, ozone (in the lower atmosphere, rather than in the stratosphere, where another set of industrially generated trace gases, the chlorofluorocarbons, continue to reduce its levels). Most of these do not stay long in the atmosphere, but are quickly washed out, albeit in the form of acid rain; this has killed forests and lake fish populations here and there, but that signal is likely to be ephemeral and hard to pick up in the future geological record.

But with carbon dioxide, things might be different. This is despite it being a trace gas, at a fraction of a per cent by volume. By contrast, levels in the early Precambrian likely exceeded 20 per cent, and, in the greenhouse Cretaceous world, they were perhaps somewhere around half a per cent, or around twenty times today's levels.

If we move closer to the present, the million-year ice-core record is unambiguous: over that time, carbon dioxide has stayed as a trace gas, showing regular fluctuations that closely match with climatic changes, while their maxima and minima, at around 280 and 200 p.p.m. respectively, show great similarities between successive climatic phases. CO_2 levels in the atmosphere are now about 380 p.p.m., and by the end of the century will likely be over 500 p.p.m., thus roughly double their 'normal' value. Now, this is still a trace amount. It will not mean that we will be suffering from lack of oxygen. But, geologically it is unprecedented, at least as Ice Age history goes, and it has taken place extremely quickly. The possible consequences for climate are now evident, and a matter of public debate. But how about the simple chemistry?

To look at that we need a little context, and some numbers. How much is there of 'our' carbon and how does that fit into the overall budget of carbon in the surface environment? Well, human activity, mainly the burning of fossil fuels and the clearing of forests, each year, currently emits about

7 billion tons of carbon into the atmosphere (that is carbon alone: if one adds the oxygen atoms to make the CO_2 this translates into about 25 billion tons). This is pouring into an atmosphere which now contains about 600 billion tons of carbon. About half of it stays in the atmosphere to make up the one-and-a-half p.p.m. that we add annually; a small part of the other half is taken up by extra plant growth in parts of the world where forests are currently expanding, such as the great northern forests of Russia.

This 'plant uptake' is itself a small part of a much bigger equation, in which land vegetation (total carbon storage about 850 billion tons) and soil (total carbon storage in excess of 1 000 billion tons) each year exchanges about 60 billion tons of carbon with the atmosphere via the twin, opposite processes of photosynthesis and decay plus respiration.

Most of the 'other half' of our emitted carbon dioxide simply dissolves in the sea, which is a gigantic reservoir of carbon, containing over 38 000 billion tons of it. There is, though, a price to pay. The carbon dioxide, dissolving in the oceans, forms carbonic acid. Some of this acid dissociates in the water to produce bicarbonate ions and hydrogen ions, which are simply protons. The protons then react with carbonate ions in the water to produce yet more bicarbonate ions. This is bad news for the many animals and plants in the water that need to use the dissolved carbonate content of seawater to build their shells and skeletons of calcium carbonate. Shell- and skeleton-building then become more difficult. Eventually—as the acid levels build up—they become impossible.

At a global level, the last two hundred years have seen the injection of a pulse of carbon into the atmosphere–ocean system that already totals of the order of several hundred billion extra tons, a figure that looks likely to be trebled, at least, by the time we run out of coal and oil. This pulse has been injected with great rapidity—it has happened far more quickly than did the changes between glacial and interglacial phases of the Ice Age. And it will take a little while to work its way through the system, not least because the ocean waters mix very slowly. The surface waters are already being measurably more acidified, but it will take at least a thousand years for the carbon dioxide to spread throughout the mass of ocean water. So, what might the consequences be?

The surface acidification is pushing the balance from carbonate to bicarbonate ions in the seawater, and is making it harder for organisms to extract calcium carbonate to build their skeletons. One can look at both the past and the present to judge the effects. At the height of the last glaciation, with less carbon dioxide in the air, the oceans were a little less acid, at least at their surface (deeper down, the situation may have been a little more complicated). Skeletons of calcium carbonate-secreting microplankton have been extracted from layers of sea floor mud deposited at that time; weighing them has shown that those shells were significantly thicker—by up to a few tens of per cent—than the same species of organism constructed today. The same organisms in the acid seas of the near future will, it seems, construct yet more fragile shells, and some might not be able to make shells at all.

At present, there are concerns as to how, in particular, reef-forming corals will cope with the new chemistry of the sea. Projecting into the future, when carbon dioxide levels in the atmosphere look set to double, it has been estimated that the production of reef limestone may decrease, by up to 30 per cent. Because reefs themselves have a kind of profit–loss balance, with production of limestone rock needing to offset reef erosion by storms and reef-coral predators, that in itself may be enough to cause reefs to shrink rather than grow. This can only add to their current difficulties.

There is a further ramification, albeit a somewhat counter-intuitive one. Surprisingly, the secretion of calcium carbonate by organisms releases carbon dioxide in the process of calcification. When a molecule of calcium carbonate is created, then one might think that this means that one atom of carbon is locked away from the atmosphere. And so it is. But only on very long timescales, of many thousands of years. At those scales, calcium carbonate—limestone—represents a gigantic, literally life-saving store of locked-away carbon. Without it, the Earth would likely have an atmosphere comparable to that on Venus, with its crushingly thick carbon dioxide atmosphere and surface temperatures of 400 degrees centigrade.

In the short term, though, in the calcification process a calcium ion combines not with a carbonate ion, but with two bicarbonate ions. This produces a molecule of calcium carbonate, and also leftovers in the shape of a molecule of water and one of carbon dioxide—the latter being redissolved into the ocean water or released back to the atmosphere. Reefs today are thought to supply, through their growth, some tens of millions of tons of carbon to the atmosphere every year. That is a significant amount—though tiny by comparison with humanity's seven billion tons.

So, more fragile sea-shells and eroded reefs may be another tangible result of humanity's global activities. But the effects of the altered atmosphere will go further. Over the next few centuries, the oceans will likely take up most—say three-quarters—of humanity's extra carbon dioxide, as that is slowly carried deeper by the slow-moving ocean currents. To be neutralized, this will begin to dissolve calcium carbonate skeletons where these have already accumulated as layers of ooze on the sea floor. It will continue to do this over several thousand years, until a new chemical equilibrium is re-established, partly by this dissolution, and partly by the effects of weathering on land. And so, when the settling of carbonate skeletons resumes, this will be on a sea floor which has been dissolved away to a depth of perhaps tens of centimetres.

There will be a gap in the record of deep-sea carbonate oozes, a widespread marker horizon, a hiatus in sedimentation, in the future limestone strata of the deep sea floor. Subtle, yes, but distinctive. Exactly such a layer was formed during the geologically brief warming event that took place some 55 million years ago, around the beginning of the Eocene Epoch, when the Earth's carbon budget was disrupted by a sudden (*c.* 10 000 years) influx of carbon dioxide, which most likely entered the atmosphere as a gigantic outburst of methane, before being oxidized to the longer-lasting CO_2. This carbon dioxide, as well as raising temperatures by between 5 and 10 degrees centigrade globally, dissolved into the ocean, and created a marked acidification event, dissolving calcium carbonate particles on the sea floor over some 100 000 years. All that remains is a layer of insoluble clay, a few tens of centimetres thick; it is quite distinctive, when seen in cores drilled from the ocean floor. The amount of carbon released (estimated at some two thousand billion tons) is on a par with that which we can release by burning all our fossil fuel reserves (roughly four thousand billion tons). Our own carbon release event is being accomplished very quickly, in just a few centuries. The ensuing acidification event must just as surely produce a similar dissolution layer on the sea floor.

The resulting deep-ocean clay layer will be geologically temporary by comparison with some of the other phenomena we have described. And, lying beneath the deep-ocean floor, it will be relatively inaccessible. Ultimately, it is bound for destruction as the ocean floor is destroyed by sliding (being subducted) into the Earth's mantle at ocean trenches. After a hundred million years, there will still be some of today's ocean floor left, accessible by drilling, just as we drill through one-hundred-million-year-old ocean floor deposits today. After two hundred million, though, very little will be left—perhaps some stratal remnants scraped off the ocean floor on to continental margins at subduction zones. There will be a wider impact, though, of this acidification event on all calcium carbonate-secreting marine organisms, including those of the plankton (such as the pteropods or 'sea butterflies', that seem particularly sensitive to such chemical disturbance) and of the shallow seas. This will undoubtedly leave a mark in the longer-term fossil record.

What of the future atmosphere? After the neutralization process, a few per cent of the fossil fuel-derived carbon dioxide will still be left in the atmosphere (because the equilibrium has been shifted in the direction of the perturbation); this means that CO_2 levels might then 'recover' to, say, 350 rather than 280 p.p.m., with a very long-term impact on climate. It will likely take hundreds of thousands of years for a further long-term equilibrium to be established, by means of further weathering on land to establish a new steady state.

That is a fair legacy, using the word 'fair' pertaining to scale rather than moral judgement. But it assumes evolution of the carbon dioxide pulse through a more or less stable oceanic system. That might not be so.

One of the puzzles of the Ice Ages has been the regularly paced variations in carbon dioxide, oscillating from a little under 200 p.p.m. during glacial maxima to around 280 p.p.m. during the warm (pre-human) interglacial phases. Not the regularity: that, as we have seen, is under the overall control by astronomical factors, slight changes in the Earth's orbit and spin, the Milankovitch cycles. The problem is, where did a few hundred billion tons of carbon dioxide go to during the glacial phases? It could not have been to terrestrial vegetation, because the temperate forests shrank back as the ice sheets pushed down from the poles, while the tropical forests were also partly replaced by grasslands, as low-latitude climates became a little cooler and quite a lot drier.

Was the carbon dioxide therefore being consumed by oceanic plankton, that then fell to the sea floor and were buried in greater numbers? Well, the puzzling thing is that this does not seem to have happened either, because if anything the plankton productivity seemed to fall rather than rise overall during glacial times. This is puzzling too, because in the drier, windier regime more nutrient-rich dust would have been blown over the oceans, thus seemingly fertilizing them and increasing microplankton populations. Nevertheless, ocean productivity, especially in the Arctic and Antarctic oceans, seems to have dropped during those glacial phases, this being reflected in a reduced supply of certain biologically derived elements (such as barium) to the oozes being laid down during those phases. So if there were more nutrients coming from above—perhaps there were even fewer nutrients coming from below? If the glacial oceans possessed, in effect, a strong lid of surface water, then it would be more difficult for nutrients from deep water to be brought up to fertilize the plankton, and so the plankton populations would, overall, diminish.

It is a plausible mechanism, but how does that explain the reduced levels of carbon dioxide in the atmosphere? Here, one has to regard the ocean as an enormous store of dissolved carbon dioxide, holding within it more than fifty times as much of this gas as does the atmosphere. Deep water is particularly important here, because there is no photosynthesis to absorb the CO_2. Rather, in deep water, this gas is produced (and immediately dissolved) as sinking organic matter—dead microplankton—is oxidized. So if there is a lid of water covering a large part of the ocean, this would hinder the exchange of gas between ocean and atmosphere. Given the enormous disparity in size between these two global reservoirs of carbon, the ocean might slightly increase its capacious reserves of dissolved carbon dioxide, while the amount of carbon in the (relatively) minuscule atmospheric reservoir would drop by a quarter.

This, of course, means the *deep* ocean store. The carbon dioxide in the sunlit surface layers of the glacial ocean would reflect the lower levels of the glacial atmosphere, and so that ocean would show greater differences between deep and shallow water in its carbon dioxide content—and so in its acidity—than does today's ocean.

It is a neat story, this control of atmospheric CO_2 by an ocean which becomes more strongly layered ('stratified' the oceanographers say) during glacial phases and less so during the warm interglacials. What it does show is that the composition of the atmosphere—and hence the behaviour of the climate—is at the mercy of the enormous, complex giant of the oceans: to clumsily paraphrase James Elroy Flecker—the dragon-green, the luminous, the dark, the serpent-haunted, the *carbon-rich* sea. What seems almost miraculous is the regularity of the (as yet undiscovered) simple physical oceanic mechanism that would thus link a glacial climate with increased stratification of the high-latitude seas.

Humanity has already changed the ocean–atmosphere carbon dioxide balance by almost exactly 100 p.p.m. from 'natural' levels: that is, by more than the total difference between pre-human glacial and interglacial times—and promises to multiply this several-fold before it is done with fossil fuels. The effect of this considerable injection can be modelled through the future oceans. But, this simple model cannot take into account the kind of knock-on effects that are almost certain to result from the kind of perturbation of the ocean-climate system that humanity has set into motion. In particular, the knock-on effects will almost certainly change the pattern of ocean currents, as a changing climate slowly alters ocean temperature structure, rather as a tugboat gradually changes the direction of an ocean liner; and as, say, melting ice alters the salinity, and hence density, of polar seawater.

Changing ocean currents will mean changed climates, regionally and globally. In this context, too, they will mean a changed pattern of carbon dioxide exchange between the deep ocean and the atmosphere. Will this mean that the deep oceans will, ultimately, absorb more CO_2 from the atmosphere? This would be a negative feedback, helping to hold back CO_2 levels and the greenhouse effect. Or will the oceans release a further tiny fraction of their enormous CO_2 store to the atmosphere, to ratchet up atmospheric CO_2 levels, and hence global temperatures, yet further? We are probably, here, in what may be termed a field for unconfined speculation. Perhaps the only certainty is that something will happen, and that the greenhouse world will throw up surprises. The status quo will not be an option.

We are in the realm of feedbacks, and might go a little further, because we are now, as an intricately interlinked earth–ocean–climate system, heading into the unknown, and a good deal of the uncertainty concerns those feedbacks. A positive feedback would ensue if the warming were to release the large stores of methane held in high Arctic permafrost terrain (and the permafrost has been shrinking back over the past couple of decades, with many of the myriad lakes at the permafrost margins disappearing as the ground ice melts, allowing the lake water to seep away into the ground). Warming might also release the equally large stores of methane that have accumulated within deep ocean sediments as 'clathrate', a waxy substance originally derived from the decay of organic matter. Warm up some clathrate and it begins to fizz, releasing methane. Wholesale undersea fizzing, precipitated by a degree or two rise in temperature of the deep ocean waters, would herald yet more warming. Then there are the huge carbon stores in the world's soils, which might supply carbon dioxide to the atmosphere, if increased temperatures were to increase the rates of decay and respiration of soil organic matter. Warming will also likely release more water vapour into the atmosphere; this is a powerful greenhouse gas in its own right, and likely to amplify the effects of increased carbon dioxide. If these positive feedbacks swing into action, the world of the Jurassic might not be so very far away.

There are negative feedbacks as well. In a warmer world with more acidic rainwater, weathering of solid rock exposed at the terrestrial land surface would proceed more quickly, helping to neutralize that acid. If much of the water vapour released by increased temperatures condenses to form an increased cloud cover, this would cut down the amount of sunlight reaching the Earth and so slow down the rate of global warming. In so far as we understand the lessons of the geological past, it is that positive feedbacks kick in first and more quickly (over thousands of years); then the negative feedbacks rein in the warming, chiefly through absorption of CO_2 by rock weathering, but only over timescales of hundreds of thousands of years.

We as humans will have an interesting time ahead. In the longer term, the knock-on effects of slight changes in atmospheric composition will ramify through the environment and imprint a signal in the strata. For any interstellar voyager versed in comparative planetary chemistry (a necessary qualification, one imagines), this will form part of the evidence of a quite singular Earthly perturbation.

THE MUD BLANKET

Coral reefs and carbonate platforms, in life and in death, are fine when you find them. But they tend to occupy only one or two per cent of the sea floor at any one time. Maybe, one hundred million years on, most of today's most striking examples—the Australian Barrier Reef, the Florida Keys, the Red Sea reefs—will still be buried a kilometre or two below the Earth's surface, or perhaps they will have been tectonically lifted into the realm of erosion, and, broken into numberless scattered fragments by the relentless grinding of wind and water, will have vanished forever. And the acid-dissolved sea floor oozes will be best seen, in that thin layer, buried deep in those fragments of today's ocean floor that survive a hundred million years' worth of subduction. It might take our interstellar explorers just a little time to find this tell-tale dissolution layer, though the catastrophic effects on calcium carbonate-secreting organisms should provide a *much* clearer signal in the fossil record.

So, will ordinary rock strata elsewhere give a sign, one hundred million years on, of the (geologically) brief but eventful passage of the human race? Perhaps they will. Let us say civilization survives until it uses up most of its oil and coal reserves. Give us this century and we will do for the oil, and make a serious dent in the coal. It is unlikely that the world's countries will agree among themselves to stem their use, arguing that to do so would trash their economies, or at the very least threaten their competitive advantage over their neighbours and rivals. Levels of CO_2 in the atmosphere will then be at least double the present values, and climbing still. Global warming by then may not be merely a couple of degrees, but might be, say, twice that, or more. As we have seen, there are geological precedents for that.

How much ice will melt? We do not know. There are too many feedback loops to predict anything with certainty. It *might* be that the world's ice caps will grow, and hence sea level will actually fall. For global warming will almost certainly cause increased evaporation from the oceans, and the water vapour so produced might simply drift over Greenland, Antarctica, and Siberia, to fall as snow, and thicken the major global ice caps, at the expense of the water in the world's oceans. That would certainly be the most comfortable option for the nine billion or so people projected to inhabit the planet by then. Currently, the interior of the Antarctica ice sheet is growing because of increased snowfall. But, as a whole, the continent is losing ice, as glaciers are flowing more quickly into the sea, particularly in its warmest part, the West Antarctic peninsula. We have to wait, and watch.

A consensus until recently has been that a doubling of CO_2 levels by the end of the century will produce a sea level rise of somewhere between half a metre and a metre. This is problematic enough for the populations of Holland, and Bangladesh, and the Maldives, but nothing directly, catastrophically to affect most of us.

But, with the atmosphere beginning to resemble that of the Jurassic, and with a biosphere in no shape to stabilize the carbon cycle (let alone that of nitrogen, or phosphorus, or sulphur), something might just go horribly wrong. The devil here lies in the detail of the dynamic behaviour of the great ice sheets, something which the mathematical modellers find so difficult to fit into their calculations. The ocean-ice system may well behave in a non-linear fashion; that is, the ice sheets may at first show little or no response to a gradual warming, or may even grow for a little time. Then, at some point which is difficult or impossible to predict, they might react very quickly indeed. Thus, parts of the Greenland and Antarctica ice sheets could well cross a threshold of stability and, over an interval which may be as short as a few decades, deflate like a collapsing balloon to shed ice into the ocean.

So let us assume a sea level rise of, say, 20 metres, a few centuries hence. Ridiculous? Remember that sea level, back in the Jurassic and Cretaceous periods, was of the order of 100 metres higher than it is now. Sea level rises of 20 metres are the small change of geological history. It is a *reasonable* assumption (the world has worked like that in the past, with rather less provocation than the indignities we as a species have heaped upon it recently) rather than an analytical one (let us make a mathematical model of the world, and calculate what would happen if we alter the parameters).

A rise of twenty metres would roughly approximate to the sea level some three million years ago, in the Pliocene Epoch, before the beginning of the Ice Ages. This is probably a conservative mid- to long-term projection for global warming. The world was warmer then, but not, by some way, as warm as in the global hothouse of the Jurassic and Cretaceous periods. Antarctica was likely a little less ice-covered than today (there is currently intense study into how much 'a little less' is) but Greenland in the Pliocene has been reconstructed as almost entirely ice-free, with only a small ice cap covering a range of mountains in the north-east of that landmass.

The nature of the resulting flooded landscape can be predicted quite accurately. Just take a map of any country and trace out the 20-metre contour. That would be the new geography, give or take the rearrangement of coastal sandbanks and deltas. For human society, there would be consequences. But here one is concerned with the wider signal: the signal to any halfway decent visiting geologist of the far future that an abrupt rise in sea level has taken place.

Sea level rises and falls on that scale (and greater) are commonplace, ten a penny in the Earth's stratal record. That is why it is not a particularly radical suggestion, especially given the heroic scale of humanity's collective re-engineering of the Earth's major carbon pathways. And if the effects stopped at the mere drowning of some low-lying coastal plains, one would not consider it greatly here. But what has become increasingly clear in our understanding of global patterns of sedimentation is that when sea level rises, the effects are felt pretty well throughout the ocean, and affect the nature of the strata being formed, just about everywhere.

Let us take a shallow sea such as, say, the North Sea between England and Holland. At present it is just the right size and shape to be dominated by powerful tidal currents, which sweep the sediment of the sea floor into the famously shifting, treacherous sandbanks off the coast of eastern England. Now have the sea level rise by twenty metres. Central London and its surroundings are flooded. Battersea and Staines disappear, while a new shoreline forms north of Egham and south of Kensington. Heathrow Airport will remain functional—just. The North Sea will grow broader and deeper, and the tides will likely dissipate and weaken within this newly enlarged volume of water. Those areas of the sea floor that had been scoured by powerful currents will be suddenly calmer. Large areas of the once treacherously mobile sandbanks will be stilled, and entombed in fine mud, drifting down from the still waters.

Something like this happened about 50 million years ago around the North Sea. Sea level rose, and a thick mud blanket covered all of the North Sea, which then stretched over what is now eastern England. This mud blanket is now preserved as the London Clay. If you travel to work through the London underground tube network, it is likely that you are hurtling past it, and, if you are a gardener, you may well be cursing it for the heavy soils it produces. It is useful to us, though, in many ways. Present as a thick layer under the North Sea, it forms a cap over the great Forties oil fields, which has prevented the oil leaking out these past 50 million years or so until, that is, humans came along and pumped it out in the later part of the twentieth century. It is a very distinctive rock stratum, a geological marker layer.

Let us go deeper in the ocean, to the end of the line for much of the world's sediment. Consider the deep ocean seaward of the mouth of the Amazon, which is not the longest river in the world, but is by some way the Earth's most massive. Just by itself, the Amazon accounts for around a fifth of the Earth's river flow. What happens to the mud and sand it carries, eroded from half of South America? Most of it gets dumped, temporarily, on the shallow sea floor seaward of the river mouth. But there is a lot of sediment continuously being carried down by the river, and the shallow sea of the continental shelf at this point is not very wide. Farther oceanwards, there is a submarine slope stretching all the way to the ocean floor of the south Atlantic, some five kilometres lower.

One wild day, the dumped mass of mud and sand around the coast may be stirred by the pounding waves of a hundred-year storm. At other times, it may be shaken by a powerful earthquake. Thus destabilized, it begins to slip and cascade down as billion-ton masses of slurry, absorbing seawater as it travels, transforming into a dense, billowing flow of suspended mud and sand, picking up speed until it is travelling as fast as an express train, all the way to the bottom of the Atlantic Ocean. And still the mass of turbulent sediment-laden water will travel, for hundreds of kilometres, like a sudden hurricane sweeping across the sea floor, until, slowed by friction, it gradually decelerates. The sediment is then released from the moving current, to form a carpet of sand and mud, maybe a metre or more thick, over the ocean floor. Repeat this at intervals, and, geologically soon, a layer-cake of sediment, hundreds or even thousands of metres thick, can accumulate off the coast of South America. This builds up the phenomenon described earlier: a turbidite fan, the product of repeated, catastrophic turbidity currents.

Turbidite fans are the Earth's garbage dumps for sediment in the deep sea, and occur off almost all the major river mouths. They have a high potential for long-term preservation. Fossilized examples, thrown up on to land by the forces unleashed during mountain-building, make up much of the Rockies and Appalachians, the Andes, the Himalayas, the mountains of Wales and Scotland. Indeed, it is hard to find a mountain range in which such strata are not common.

Turbidite fans, though mostly forming on the deep sea floor, are sensitive to relatively slight changes in sea level in the sunlit waters thousands of metres above them. For, imagine a drop in sea level of 50 metres. What had been the shallow sea floor of the continental shelf becomes land, to be traversed, and eroded, by rivers that now reach almost to the edge of the continental shelf. Fifteen thousand years ago, when sea level was lowered like this, and the last glaciation was at its height, the Amazon River, longer, and carrying much more sediment than now, was almost continuously pouring vast amounts of sediment on to the Amazon turbidite fan. Today, the fan is relatively quiescent by comparison, and is growing only slowly. If sea level was to rise still further, then the Amazon would be shortened dramatically (for the Amazon Basin is almost flat, and not far above sea level), and the diminished amount of sediment it would carry would largely be trapped around the new shoreline, far from the edge of the continental shelf. The mighty Amazon turbidite fan would receive few or none of the turbulent hurricanes of mud and sand, and would virtually cease growing upwards.

Thus, changes in sea level are not only felt and recorded around the shoreline, but the effects ripple out across much of the sea floor. Those effects, translated into different types of sedimentary strata, would be among the most obvious effects to be observed by a geologist, whether human or not. In the case of the Amazon fan, our putative sea level rise would see relatively thick, rapidly deposited sedimentary strata succeeded by thin, slowly accumulated sediment layers. If an area of muddy sea floor receives virtually no sediment, then slow-acting chemical changes lead to a thin, but highly distinctive layer of manganese, and phosphates, and distinctive greenish iron silicates, these minerals being the result of the interaction of seawater with a stable, inactive sea floor over a long period of time. It is a signal that something happened to sea level.

Such a layer is known to geologists as a maximum flooding surface. It is part of the terminology of a branch of geology, termed sequence stratigraphy, that matches up stratal successions by exploiting patterns in the strata produced across the world by changes in sea level. Ironically, sequence stratigraphy is particularly practised by oil geologists to predict the location of oil-bearing strata. A maximum flooding surface may be the first sign of our presence that future geologists will notice as they start surveying the strata of our planet in search, perhaps, for resources to power their own colonization.

There may be still further ramifications, for it is surprising what the long-term consequences can be as climate and sea level begin to change. There is some evidence that coastal volcanoes may erupt more frequently during episodes of sea level change, as the weight of the extra seawater presses on to shallow magma chambers, or as underground magma comes into contact with percolating seawater, generating steam explosions that may trigger larger eruptions.

Even mountains can be moved. For instance, the Andes are in general formed by a single mechanism: the Pacific plate sliding under the South American continent and buckling its western rim. Yet the mountains of the middle part of the Andes are about twice as high as those to north and south, reaching six kilometres above sea level. Something is holding them up, and it has been suggested that this is essentially a response to a cold Antarctic current impinging upon the coast. This in turn creates arid conditions over the adjacent coast and mountains. Hence there is little river flow, and so little eroded detritus is brought into the oceanic trench. With little sediment to lubricate the downgoing Pacific plate, this then forces the mountains higher up. The implication is that if the future climate becomes wetter and the trench fills with sediment, over millions of years the newly greased subduction zone will let the middle Andes subside to a more 'normal' level. A mighty effect from a tiny beginning, if so! But also a mightily subtle one, in which cause-and-effect would be hard to deduce in any far future analysis.

EPOCH, PERIOD, ERA?

Let us summarize the wider signals of our brief sojourn on the Earth. The signals probably first became apparent 10 000 years ago, when we became the prime candidate for the extirpation of half of the world's large mammal species. That is a rather minor event, likely to be lost amid the noise of the world's happenings. It would take a dedicated and patient vertebrate palaeontologist of the future to discern it. For most of the next ten thousand years, changes are just local: the deforestation of the Mediterranean landscapes, and then those of northern Europe. Three hundred years ago, began the unprecedented acceleration of both the numbers of humans on the planet and of their individual impact. Deforestation takes place over all over the planet, as virtually every part of the landscape amenable to cultivation becomes cultivated. Then global chemical changes, as several hundred million years' worth of accumulated carbon were extracted from sedimentary strata and, in a few centuries, injected into the atmosphere. More widespread extinctions, global warming and a sea level rise. One might begin to draw parallels, for the beginning of the Jurassic was not unlike this, with a sudden, global sea level rise, a striking change in the nature of rock strata being laid down over most of the world, whether on land or sea, and a revolution in the Earth's life-forms.

It has already been suggested that humanity's alteration of basic geological and evolutionary processes means that we are already living in a quite new interval of geological time. This interval has been called—at first informally and perhaps half-jokingly, but now more seriously—the Anthropocene by some scientists (a term introduced by Paul Crutzen, not a geologist, but a Nobel Prize-winning atmospheric chemist), and a reasonable starting point for it would be around the beginning of the Industrial Revolution. It may, perhaps, be an entirely reasonable interpretation of events. The only question that remains is to ponder the scale of this geological change. For, there is a hierarchy of geological time units, just as we divide more recent time into centuries, years, months, and days; or, perhaps a closer analogy would be with the reigns of individual monarchs, say Henry VIII, and their location within dynasties of kings and queens, such as the Tudor Dynasty.

Thus the Jurassic Period forms part of the Mesozoic Era (roughly speaking, the 'Age of Reptiles') which in turn is packaged together with a couple of other eras into the half-billion-year span of the Phanerozoic Eon, which started when multicellular life appeared, and continues today. The Jurassic Period has been split up into the smaller time slices we have mentioned above, such as the Early Jurassic Epoch, and in turn split into the yet smaller-scale ages, such as the Toarcian Age, and so on.

So, what is the Anthropocene turning out to be? An age or epoch? Will it develop into a period or even an era? Or (heaven help us all if that happens) an eon? Formally, we are now living in the Holocene Epoch (which started some eleven thousand years ago, when the ice last receded), which is the younger part of the Quaternary Period, which started some 2 million years ago (formally, its base is placed at 1.81 million years ago) when the Ice Ages took a real grip on the planet, and that in turn is part of the Cenozoic Era (generally corresponding to the 'Age of Mammals', an era which is now taken to start at the instant of impact of the meteorite which, 65 million years ago, cut off the dinosaurs in their prime).

We cannot yet predict the scale, or persistence, of the changes we are ushering in, for they are just beginning. There yet survives a reasonable proportion of the species that were around us when *Homo sapiens* first evolved; coral reefs, though sickly, are still functioning in some fashion; global temperature has not yet gone up by much more than half a degree; and the coming marine transgression has hardly begun.

Nearly three decades ago, the oceanographer Wallace Broecker suggested that global warming would lead to a brief 'super-interglacial' interval, after which, some thousands of years later, the normal course of the Ice Ages might resume. That's a forecast that was remarkably prescient for its time, but it strikes one now as conservative. Geologically, that would be a blip. The Anthropocene would take its place as a new epoch of the Quaternary Period, a modest footprint of our activities.

What, though, if the extinction event now beginning to unfold encompasses up to, say, half of the world's species. And what if our carbon glut sets off some serious positive feedback effects, such as warming the ocean floors and continental permafrost regions to trigger gigantic outbursts of methane into the atmosphere, to turn up the heat even more? We would almost certainly lose most of the Earth's ice caps, return to a Jurassic-like hothouse world, and submerge most of the continents in a spectacular marine transgression. It might then be hard to grow back the ice caps, even over geological timescales, and even if humans, perforce, have largely disappeared from the scene.

The Anthropocene would then be, at the least, a period-level change, rivalling that which ushered in the Jurassic. Perhaps, if we make enough of a mess of the world, we might compete with the Yucatan meteorite, or with the mysterious forces that, almost exactly a quarter of a billion years ago, suffocated most of the Earth's oceans and killed off an estimated 95 per cent of the world's species, bringing the Palaeozoic Era to a dead halt.

Science fiction? Would that it were, even if such changes would guarantee a superb fossil record for the human race and for all its earthly monuments. But one would lay odds, right now, on the Anthropocene attaining the status of, at the least, a geological period. For those who are not completely hung up on our everlasting immortality, a reasonable interim aim might be to try very, very hard for the Anthropocene to develop into something no greater than an epoch-scale event. That might, just, save a few billion human lives.

Whatever the scale of humanity's ripple effects, it might not be immediately clear that they had been caused by a home-grown, more or less intelligent civilization. Mass extinctions, sea level rises, carbon dioxide outbursts—these have all happened before in Earth's history, due to entirely natural causes. The merry-go-round migrations before the extinctions, the pollen from the mass-produced cereals—well, those would appear strange. But, the Victorian-age geologists proposed networks of land bridges to explain the occurrence of very similar dinosaurs on very different continents, before there came the realization that the continents themselves, on which those dinosaurs had lived, had drifted apart and coalesced into new patterns.

One can imagine our future chroniclers becoming more and more aware that this particular event was out of the ordinary, and dreaming up hypotheses that would explain all those odd and puzzling details. They would search for evidence—and eventually find it. Somewhere the hundred-million layer will be both at the surface *and* will happen to be of, say, a city-bearing delta. For the future palaeontologists, that would be a spectacular find, akin to Walcott's discovery of the Burgess Shale a century ago. And then, the real fun would start.

Traces

We can talk about little else but the new discovery. Perhaps it should not be such a surprise, given the complexity of life on this planet. And we have seen indications previously, but could not be sure; the form of petrifactions here are so diverse, and often so enigmatic. Yet there is no doubt now. An organized culture appeared, or arose, and settled on the land surface. We have little detail yet, and are now excavating. It seems to have been extremely short-lived. The site would likely not have been discovered at all had it not been associated with one of the perturbation events that we have been trying to decipher.

MAKING YOUR MARK

What fossils did you make today? If you imagine a fossil to be a dinosaur skeleton in the grand entrance of a museum, you might think that you would not have the opportunity to become a fossil until the day of your funeral. However, your fossil-making capacity is far greater. You may, for instance, be making a contribution to the potential fossil record each Wednesday evening, as you leave the wheelie-bin out at the end of your drive. Humans have the capacity to make fossils all the time: each time, for example, that one defecates, or walks through the park.

Evidence of past life comes in two main varieties: the actual bodily remains of a once-living organism and any traces left of the activities performed by that organism. Humans, clearly, produce both types, termed 'body' and 'trace' fossils respectively. Future palaeontologists, though, trying to characterize life in the Human Period will undoubtedly produce an incomplete and perhaps hugely misleading reconstruction. The fossil record of human beings, like the fossil record that we study today, has inherent biases.

How will future palaeontologists categorize human trace fossils? It is first necessary to emphasize that there will not be one type of human trace fossil, because humans have not one but many types of behaviour. This is a first principle of trace fossils: a single type of organism can make many different traces.

Let us choose an example and follow him through the course of a day. We may call him Robinson. Robinson is on a cruise when—catastrophe!—his boat capsizes in a storm and he washes up on a deserted island. He crawls up on the beach, and lies there for a while, catching his breath, leaving an imprint of his body where he lay on the sand. He walks around a little on the beach, leaving a trail of footprints (he lost his shoes in the swim for the island), looking for food. Robinson remembers that beaches are a good place to find shellfish. He decides on this as a strategy for lunch. So, he digs up some clams, each time leaving a little hole in the sand and a little pile of sediment next to the hole. He makes another pit, which he fills with stones and in which he makes a fire. He bakes the clams, sits down to eat them, and, as he finishes each one, he throws it over his shoulder to form a mound of the leftover clam shells. But, he is tired by the hot sun, so he sits down in the shade of a palm tree to rest, leaning up against the trunk, hands crossed over his chest. He realizes after a few minutes, though, the shadow of the tree is moving with the sun, and he moves over a few inches to compensate. He does this for a few more hours. Robinson then decides to go for a wander around the island, just to see if there is anything interesting that he might be missing. At one point he crosses a mud flat, but the mud is very gooey and he begins to sink into it. It takes all his strength to pull his feet out of the mud. In fact, the episode scares him so much that he has to disappear behind a bush for a while. Robinson begins to look for some shelter for the night and finds a (luckily) uninhabited cave. He falls asleep peacefully in this new home and dreams of rescue and of Jeannie.

Now, Robinson has only performed the most basic of animal behaviours: resting, moving, eating, defecating, finding shelter, and escaping from danger (not necessarily in that order). Even an average worm does that on a daily basis. But think about how many different types of traces Robinson made and how complicated some of them were. For example, he left at least four kinds of resting traces: when he lay on the beach, when he sat on the beach for lunch, when he sat up against the tree, and when he lay down to sleep. All of these will look different, depending upon which parts of his body were in contact with the sediment. Lying down on the rubbly floor of the cave, moreover, will leave a different type of lying-down trace than lying down on a sandy beach. This example illustrates a second principle of trace fossils: the same structure may look different when formed in different types of sediment.

Next, let us watch a day in the life of Farmer Archer. Farmer Archer wakes up in his fifteenth-century farmhouse to the sound of the rooster crowing. He cooks some eggs for breakfast, freshly collected from the hen house. Farmer Archer and his dog Lassie head off to the cabbage field, which Farmer Archer ploughs into furrows and then plants out with cabbage seeds. He eats a sandwich out in the field. At some point in the afternoon, Farmer Archer decides to dig a small hole to check the dryness of the soil, and Lassie, as dogs do, decides to dig a hole right next to his. Then, come evening he returns home to a dinner of fried chicken and passes the time carving a wooden cradle for his sister's baby, due to arrive in June.

It has been a fairly normal, even humdrum, day in the life of Farmer Archer. However, he has been quite active enough to illustrate for us a few more principles of trace fossils. First, one may consider his ploughed furrows. On first inspection, we might say that these are a strictly human trace. But, consider all the earthworms crawling through the soil, all the larvae making their homes in that soil, all the birds walking over the tops of the furrows looking for the cabbage seeds and the worms, and all the root structures that the budding cabbages eventually send out. The ploughed furrow, here, turns out to be one structure made by many different kinds of organisms. Thus, a third principle of trace fossils emerges.

Farmer Archer and Lassie also illustrate a fourth principle of trace fossils: different types of organisms can make similar traces. Whilst, clearly, Farmer Archer and Lassie will make completely different types of footprints, they may dig exceptionally similar types of holes. Perhaps a better example comes from the fossil record. There is a trace fossil called *Skolithos*, a simple, vertical, drinking-straw-shaped burrow, found in rocks dating from the Precambrian to the present day, a span in excess of half a billion years. Biologists have observed three different types of worms, a number of fishes, sea anemones, and a suite of arthropods making burrows that, fossilized, would be called *Skolithos*. So, while trace fossils are given genus and species names, just like living things or body fossils, they do not necessarily relate to a single biological species.

Trace fossils, consequently, are not classified in the same way that we classify organisms, on the basis of the closeness of their relationships to other living organisms, but on the type of behaviour displayed. Between Robinson and Farmer Archer, we have observed most of the major types of trace fossils. There are resting traces: this is Robinson sitting and lying down, while Robinson when crawling and walking is making fine locomotion traces. Robinson's habitation of the cave is, loosely speaking, a dwelling trace. Those who study trace fossils—ichnologists—make a distinction between traces of habitation made in the sediment or rocks as opposed to those built on top of the sediment, which are called edifices. Farmer Archer, for instance, lives in a fine edifice, of which he is inordinately proud.

Feeding traces are common in the rock record, and distinctions are made between different types of feeding behaviour. A three-fold division is often made, between things that eat the sediment, predators, and grazers. Here we are in something of a quandary. Does Farmer Archer pulling up his carrots count as producing grazing structures? There are gardening traces, and here we have Farmer Archer planting his cabbages. This may seem like something that is unique to human beings. But leaf-cutter ants, for instance, have evolved sophisticated behavioural patterns to farm the fungi that they feed on. A comparable fossil example may be a strange hexagonal network of traces called *Palaeodictyon*, which can adorn the surface of certain strata formed in ancient deep seas. The unknown creature that made this seems to have been going repeatedly over the same ground, collecting or growing food in the troughs and coming back for it again and again.

Escape structures are common in rock strata, representing those times when some unfortunate worms or molluscs, quietly going about their daily business, were suddenly buried by a layer of choking mud, as a hurricane, say, swept through above them. Those that survived were the ones that managed to dig their way back to the sediment surface. Robinson escaping from the mud flat went through a similar experience. Rather less dramatic responses to environmental changes are equilibrium structures, and here we have Robinson moving around the palm tree to keep in the shade. Typical structures of this kind in rocks are called spreiten, where organisms adjust their vertical burrows up or down as small increments of mud or sand are added to, or eroded away from, the sediment surface at which they come out to feed. There are also breeding traces. Farmer Archer built a baby cradle, for instance, while bees and wasps, showing comparable skill, build hives.

One more thing to do here is to compare Robinson's and Farmer Archer's trace fossils. In some ways, they are similar, but there are differences. Robinson's footprints, for example, are foot-shaped, while Farmer Archer's are Wellington boot-shaped. This difference illustrates a fifth principle of human trace fossils (not necessarily used by palaeontologists because it has just been improvised for our particular purpose): the same organism has different behaviours in different environments. It is rather harder to think of fossil, non-human examples, but there are a few, if one does not mind stretching a few analogies close to breaking point. The coral *Acropora*, for instance, is one of the most important components of coral reefs, both modern and ancient (though, geologically, not-so-terribly ancient). In sheltered parts of a reef it grows long, slender branches; in those areas pounded continuously by waves it forms rounded, shock-resistant colonies.

With increasing urbanization, too, it becomes more difficult to distinguish a trace made by a particular individual, and linking a particular trace with a particular individual is usually the best way to establish what made a trace. For instance, some horseshoe crabs were carried, about a hundred million years ago, into the lethal, hypersaline lagoon that was to become the Solnhofen Limestone of Southern Bavaria, famous as the site where the *Archaeopteryx* was discovered. These made a kind of final, brief death march, now seen as a series of fossilized claw-tracks on the limestone stratal surfaces, that end in the fossilized corpses of the unfortunate crabs themselves. This kind of connection is always more likely to be made with Robinson (if he had not survived his encounter with the mudflat) than with Farmer Archer.

Robinson, in his role of castaway, was essentially behaving as did the dinosaurs in the far-off Mesozoic Era. He was doing his best to survive in a cruel world, with little but his instincts and his wits to help him. He was gathering. He was hunting. He was trying to find himself, in between gathering and hunting, a quiet and comfortable place to shelter in. The tracks he was leaving are, in essence, the same as the footprints the dinosaurs left behind for us to wonder at, in the Purgatory Valley of Colorado, for instance, or scattered on sandstone surfaces on the shores of the Isle of Wight. Any individual human only has two feet to leave for posterity, but consider how many footprints one can make in the course of a lifetime. Consider too how many of us there are: six billion rising to nine billion in another 50 years' time, all being well. That is likely a biomass comparable to that of the dinosaurs at any individual time in their long history. And that represents an awful lot of trampling.

Footprints, though, are easily washed away, so only a tiny fraction of our walking traces will survive. Farmer Archer goes further in modifying the landscape, in ploughing, and in building an edifice to shelter himself and his family. There are some parallels among the dinosaurs. Occasional dinosaur nests have even been found, in the Gobi desert, for instance, or around the vertical limestone strata of that spectacular mountain, Mont St. Victoire, in the south of France, which so transfixed Cezanne that he painted it again and again. But, while Farmer Archer dug himself a pleasantly cool cellar below his farmhouse, only a single dinosaur species, as far as is known, seems to have constructed a burrow to live underground. But if we think at a grander scale, and consider the levels of construction and cooperation that human beings are capable of, our analogies may stretch much farther.

Today, few people live on desert islands. And, in the more developed countries, an ever-smaller proportion live the good life, farming in the wide countryside. Most of us live in towns and cities. And here, collectively, we have become giants. Giants, that is, as regards the production of human trace fossils, both in scale and in quantity. At an ever-increasing rate, we are currently creating examples of the most amazing (and, in the far distant future, perhaps the most puzzling) trace fossils likely to appear in the history of the planet. It is time to consider the most singular, the most various, and the most obviously *constructional* contribution of humans to Earthly geology: the Urban Stratum.

URBAN TRACES

What would be the first fossil evidence of undoubtedly intelligent organisms to be unearthed by our future chroniclers? The clinching evidence that, tangled in with this abrupt disturbance of the Earth's environment, there had been intelligent, organized, manipulative, colonial—and briefly successful—beings?

It is clear that most of the sedimentary record of the Earth is marine, and that will hold true for our own time interval—for what we may call the Anthropocene. We, on the other hand, are not marine beings, and almost all lives are spent, and our constructions are built, on land. So, will the first direct finds be of our brief sojourns into the ocean world: bottles or cans tossed overboard, or even whole shipwrecks? Possibly so. But here the evidence would be fragmentary, partial, tantalizing. Certainly not on the scale associated with the scale of environmental change to be shown by the strata.

The real evidence, the classic fossil locality-to-be, the scientific goldmine, would be when the marine strata of the Anthropocene were traced into rocks representing shallower and shallower water until, say, a fossilized delta or coastal plain was found. This would be an equivalent, say, of the delta upon which the coal swamps of the Carboniferous formed, or of that, now exhumed in the Weald of England, upon which dinosaurs left their footprints (and occasionally their bones) in the early Cretaceous. Once they hit upon such strata—not by accident (for that would be like finding a needle in a haystack) but by following, carefully, patiently, the rock layers associated with the mass extinction—our far-future explorers would be poised, finally, to discover a fossilized city.

Consider the sheer scale and present-day growth rate of these constructions. We are in the age of megacities, built ever larger as population not only expands, but also migrates from the countryside to cluster around sources of work and shelter.

What city? It might be New Orleans, or Haiphong, or Shanghai, or Amsterdam, or Venice, or Port Harcourt, or Dhaka. These are just a few of the cities and megacities that today spread across coastal plains, and over river floodplains and estuary margins. These, hence, are firmly sited on down-going tectonic escalators, the weight of the deltaic sediment that they are built on inexorably dragging them down. And all are at or just above sea level (or just below it, in some cases, behind protective walls), so making them vulnerable to drowning by even the slightest of sea level rises. Once drowned, they will be removed from the realm of erosion into the realm of sedimentation, as if placed in a pickling jar.

Manchester in Lancashire, though, San Francisco, Lhasa, Clermont-Ferrand, Quito, and La Paz are all high in the erosional realm, and mostly on upwards bound escalators. Their chance of long-term fossilization is effectively nil, however strongly they are built to withstand storm and fire and flood. Mexico City has a good short-term chance of fossilization, being built on a former lake basin next to active, ash-generating volcanoes; but its long-term chances are poor, as that basin lies on a high plateau, some two kilometres above sea level. The only ultimate traces of the fine buildings of those cities will be as eroded sand- and mud-sized particles of brick or concrete, washed by rivers into the distant sea. This is still a signal, admittedly: the material from which those cities are built cannot just disappear. But it will be a very subtle one.

Let us return, then, to the low-lying parts of the human empire, and see just what is available for fossilization. It will represent a cornucopia of palaeontological wealth to make even the Colorado valley, criss-crossed as it is with dinosaur footprints, seem small and mundane and childishly simple.

Think of what has been brought into New Orleans, say, to construct its modern skyline. An average skyscraper, say, is made of thousands of tons of concrete, plus a great deal of steel and glass, some copper, various plastics. Stone, too—polished granite slabs, perhaps imported from Scandinavia, in the entrance hall, and marble cladding.

And that is just what we see at the surface. Buried underground there is a great deal more. More concrete, in particular. Some goes to make a raft on which the whole structure sits. And even that is not enough. The ground beneath New Orleans is, to a construction engineer, as treacherous a foundation as can be found anywhere. There are sloppy, waterlogged clays, and layers of loose sand and of partially decayed vegetation. If one tries to balance a skyscraper on that kind of foundation, it would, like as not, tip over after a few years and fall to the ground. There is a lovely, much-photographed tower in Pisa that beautifully demonstrates the problems of building tall structures on soft ground. It has taken all the ingenuity of the engineers and lire of the Italian state to keep it precariously, and astonishingly, poised between stability and collapse.

The only *secure* way to anchor a spire of a few thousand tonnes of concrete, steel, and glass into the geological equivalent of blancmange is to sit it on a raft of concrete and then literally to pin down the entire mass with concrete piles driven 50 metres or more into the ground. And each skyscraper is not an island, of course: there is a complex skein of water-pipes, sewer pipes, electricity cables and gas-pipes, fibre-optic cables, ramifying through the ground to link these modern castles.

Thus, if we compare ourselves with the dinosaurs, each in our time, masters of the terrestrial realm, top-of-the-food-chain and so forth, then, at least in the business of producing traces, we come out firmly on top. This is despite the fact that we are only one species, that has been around for much less than a million years (and building cities for only a tiny fraction of that interval), while the dinosaurs represent many species, which collectively occupied the Earth for a hundred million years.

Each of us humans (in the developed world, at least) on average, in our lifetime, uses some 500 tons of sand and gravel, plus limestone, brick clay, and asphalt. From these we create our roads, houses, and foundations, schools, hospitals, restaurants, and multiplex cinemas. Together with the iron, steel, copper, plastic, this can, over the human generations and the centuries, pile up to form a great mass of material, for new houses tend to be constructed upon the rubble of the old. In urban areas the accumulated rubble of centuries of building forms a significant geological deposit. On geological maps it is often termed 'Made Ground'. For the purposes of this narrative, let us call it the Urban Stratum. In old, well-established cities it can be tens of metres thick, and in cities such as New Orleans, the pilings that keep the skyscrapers safely scraping sky additionally form a kind of subterranean inverted forest of concrete.

Our influence extends yet deeper. For cities, rather like ant and termite nests, need a constant inward supply of food and materials to maintain their survival. As regards the food, well, Farmer Archer helps as best he can. As regards warmth (and, increasingly, the provision of cooling, in summer), humans have not engineered our cities nearly as well as the termites have constructed their nests, where the constant internal temperature is a result of quite beautiful engineering, ensuring a marvellous balance of warmth (a by-product of the insects' metabolism) and ventilation.

Our rather cruder human arrangement used to be based, for centuries, on burning wood and charcoal. Then we used up the supplies of wood, as the forests were consumed. Then came coal, and we ingeniously effected the removal of entire strata, often from several hundreds of metres below the surface, the massive effort of extraction necessitating networks of shafts and tunnels far underground, prevented—temporarily—from collapsing by props of wood or steel. These are, in effect, burrows: like those made by the worms in the mudflat, only on a giant scale.

After that, there was oil, the defining economic, social, political, cultural, and military symbol of the last hundred years, the ultimate fuel of convenience, the fuel that has driven almost everything else in our lives for the past hundred years (and might do for the next fifty, if we are lucky). Its exploitation has not involved us invading the underground realm to quite the same extent that the exploitation of coal has done. Nevertheless, thousands of boreholes—iron-, concrete-, and mud-lined holes up to a metre across and up to a few kilometres deep—make each oilfield the equivalent of a giant pincushion. Now, with our civilization's rather nervous transition to nuclear energy, where has the nuclear waste to go? The least worst option is to place it deep underground, as a very distinctive gift to the geologists of the future: whatever shape these beings take, they will likely treat high levels of radiation with respect.

How well can we employ our ichnological classification to the internal structure of the Urban Stratum? Habitation traces, for a start, that are technically termed domichnia: that is easy—flats and houses. But there are also offices and factories—these do not fit quite so neatly: worms and crustaceans do not have separate living-burrows and working-burrows. As regards locomotion traces (termed repinichnia when applied to fossils), most of our urban footprints do not stand a chance of posterity, as our Gucci shoes and designer trainers pound the unyielding pavement surfaces. But instead of this, we leave a network of collective locomotion traces: pavements, roads and superhighways, railways and airport runways that extend far beyond the urban centres themselves. These are considerably more hardwearing than the deepest footprint. And what of escape traces?—well, many office workers likely think of their homeward subway or underground journey in terms of escape.

There is, though, at least one fundamentally new class of trace fossil that we have created, or evolved. One may call these the pleasure traces, or, if we are to be properly scientific about this, let us term them frivolichnia. For our species is the only one to devote a large slice of its effort and resources to the pursuit of pleasure. Think of it: cinemas, sports stadiums, parks, museums and art galleries, theatres, gardening centres, high class restaurants (while low class restaurants, naturally, only represent feeding traces, or fodinichnia). This is one aspect of our legacy that our far-future inquirers will either understand and easily fit into their interpretive framework (if they are like Douglas Adams's science fiction creations), or one they will find inexplicable (if they are like Mr. Spock).

Another class of traces has counterparts in the natural world, but of an activity that we can claim to have taken much further than have mere animals and plants. These lower orders of life kill each other for food, or sometimes for competition over a mate. The human species has carried out murder on a grand scale, and has devoted (and still devotes) a good proportion of its resources to this activity. The remains of swords, guns, bullets, warships, and the ground torn apart by bombs, landmines, and tank tracks might be classed under another newly invented term as kil-lichnia. There does not seem to be a palaeontological equivalent of this, which perhaps says something about the nature of progress.

Cities, of course, are full of *things*—our material possessions: cars, chairs, tables, pianos (acoustic and electric), tennis rackets, computers and televi-sions and CD players, microwave ovens, parrot cages (some with parrots in them), even books. Then the small fry, shed from our lifestyle in their bil-lions, much as a plant sheds pollen: ballpoint pens, for instance (think how many of those we get through in a lifetime); paper-clips; coins, keys; plastic knives and forks and spoons; plastic drinks bottles and paper cups.

The list here could go on almost forever. These objects are currently scattered amongst our houses, flats, shops, supermarkets, and warehouses (another new class of trace fossil here). Most have relatively short useful lives, though, and eventually wind up in our dustbins and skips, and thence to the municipal rubbish tips, usually old gravel-pits—which are often more profitably employed as waste dumps than ever they were as sites of aggregate extraction. Contemporary archaeologists place great store by ancient rubbish tips, or middens, for in sifting through them they can learn a great deal about daily life in, say, the Middle Ages. Contemporary society throws away infinitely more than the thrifty villagers of pre-Renais-sance times ever did, and our gigantic waste dumps, if fossilized, might overwhelm the interpretive capacities of our far-future observers.

A particular class of fossils would be represented by our clothes. These do not seem to fit easily into any category of trace fossils. Rather, these are more akin to an external armour for protection and insulation, doubling in many cases as a form of artefact for sexual display. The nearest func-tional equivalent seems to be the plumage of birds, or the exoskeleton of an arthropod such as a crab or a lobster. As in the case of the arthropods, the external covering is periodically discarded as it grows too tight to fit the growing organism, and a new one is made. As in the case of some arthropod species also—and in most birds—there is a strong sexual dimorphism in the external covering, with different patterns for the male and female of the species. Unlike the case of other species, though, there is also a marked polymorphism of this external covering between different communities and different generations of this same single species, with the sophistication, if not the functional practicality, of these human body-coverings generally correlating with the resources that individual humans dispose of. Given the uneven distribution of these resources among individual humans, one can predict that a Marks & Spencer's pullover is more likely to make it into the fossil record than an Yves St. Laurent evening creation. Together with the clothes, there are vanity fossils—the infinitely tiny fraction of artefacts that comprise the time capsules buried for discovery by future generations, a human invention that further broadens the scope of conceptual palaeontology. The various preservable components of our current Urban Stratum certainly encompass a bewildering range.

However they may be classified, it is easy to see that humanity has re-engineered substantial parts of the planet's surface, and has done so with exceptional speed. Other creatures have certainly had an effect on the planet, but nothing quite like our megacities has appeared previously. The first photosynthetic organisms, by flooding the atmosphere with oxygen two billion years ago, changed the world forever. Now *that* is impact, but it is not a city-analogue. A termite mound, conversely, can be compared to a city: its architectural complexity, heat control, and air conditioning are arguably the equal of anything that we can make. But, termite mounds cannot be said to have global impact.

There is, though, a living phenomenon that combines infinite constructional complexity and global, landscape-altering scale. Here again we meet the corals, and the quite stupendous reef structures that they build. Small be-tentacled metazoans with a talent for making limestone are, in a sense, our immediate competitors for immortality. Both skyscrapers and coral reefs are basically large masses of biologically constructed rock, worthy monuments to our respective phyla. Our immediate contribution to the skylines of New Orleans, Amsterdam, and London certainly seems impressive enough, and satisfactorily solid, as regards both the visible and invisible parts of it.

It is worth simply recalling the size and the status of coral reefs as prehuman cities. The Great Barrier Reef, stretching 2,000 kilometres along the Australian coast and covering a quarter of a million square kilometres, is justly renowned. More surprisingly, it seems to be a baby reef, less than a million years old. It may be visible from space, but it is only just beginning the process of wholesale landscape modification.

A reef *is* a biological city, quite as complex as New York, Beijing, or London, but built on a quite inhuman scale as regards both time and space. The Pyramids, Great Wall of China, and New York are striplings by comparison. Still, we humans have only been in the city-building trade for a few thousand years. The coral organisms have had hundreds of millions of years. We are catching up with impressive speed.

JOURNEY TO THE UNDERGROUND

Humanity has, with extraordinary rapidity, created abundant traces of its own activity, traces composed of brick, concrete, steel, plastics. These traces can, with a half-decent concatenation of tectonic subsidence and sea level, be buried by sediment and carried below the realm of erosion. So far, so good.

That is a beginning, certainly. But the journey ahead, to the projected rendezvous with the interstellar explorers, is much, much longer. The city-remains, carried along on the tectonic escalator, will endure these ages in an environment that is more alien to us than the depths of the sea or the chilly reaches of outer space. It is the world of the underground, a world of dark and heat and enormous pressures and the slow coursing of subterranean fluids. It is the birthplace of earthquakes and oil and natural gas. How will our cities fare, then, in Pluto's realm? Will the original shapes of the buildings and foundations, entombed in the rock strata, be recognizable?

Will they disintegrate, disappear without leaving a trace? Will they simply survive underground, unchanged? Will they change into different mineral forms? These are the questions that will determine the nature of the evidence that we will leave behind us. It might help to consider the Urban Stratum from another perspective. Urban it certainly is; but just what are its qualifications as a stratum?

One might start by taking the composition of a city, and translating it into its essentials, into geological materials. For then one can make the closest comparisons with those natural materials that have survived prolonged burial within strata—and also with those that have disintegrated, decayed, disappeared once in the underground realm. Most of the items of human manufacture are made of *artificial* materials, certainly. But, beneath their modern aspect and form, many ingredients are downright primordial. It is these that will provide the best analogues to gauge the progress of a buried city, as it is forward-modelled through many millions of years of time.

Take that main city-ingredient, concrete. It is one of the most potent symbols of the modern age. In truth, it has a longer history, being made in Roman (and perhaps ancient Egyptian) times. But it was Joseph Aspdin who can be credited with coming up with the decisive, world-conquering recipe in 1824 from the simplest of ingredients, like an accomplished chef carrying off his third Michelin star: take crushed limestone and clay; bake together, then grind into powder. To make concrete, take one part of this cement and two parts of sand and gravel, mix with water, shape more or less how you like, wait for a succession of highly complicated hydration reactions to take place, and the mixture will set hard as rock, even under water.

It is hard to make precise predictions as to what exactly will happen to this man-made rock after a hundred million years underground, particularly as there are many different types of underground. Concrete was not tested, originally, even for short-term longevity: witness the expensive problem of 'concrete cancer'—a catastrophic weathering reaction—that caused buildings and bridges to crumble in the 1960s and 1970s.

Certain components of concrete, though, have geological durability built into them. Take the main ingredient, some two-thirds by weight:

sand and gravel. Geologically, sand is defined as sedimentary particles between sixty-four thousandths of a millimetre and two millimetres in diameter. Gravel is pebbles, between 4 mm and 16 mm in diameter (above that size are cobbles and then boulders, while the rather awkward gap between 2 and 4 mm diameter is taken up by granules). That is geology; the building trade has its own definitions. Both sand and gravel are termed aggregate: fine aggregate is the term used for material below 5 mm in diameter, and coarse aggregate for particles greater than that. No matter. The exact size is not so important here. The history, though, of those particles of sand and gravel in the geological past, and the ancient journeys that they made, will provide the best idea of their likely future behaviour, as components of a deeply buried concrete block, into the far future. Sand and gravel is dug out of pits and quarries. But how did the sand and gravel get there in the first place?

The natural production lines of sand and gravel are familiar ones, in other contexts. Many of us are accustomed to spending two weeks every year at a highly efficient modern sand and gravel factory: a beach resort, something to walk across in flip-flops and appalling Bermuda shorts while carrying the latest blockbuster novel. Here, the waves lapping on that beach—or pounding on it, on stormy days—are ceaselessly busy. The waves roll sand grains and pebbles along the beach surface, while the mud and clay, suspended in the seawater, are ultimately carried farther out to sea. The sand and gravel particles, before travelling to their final resting place on the beach, are continuously under chemical attack by the water itself. Particles which are made of minerals that show any physical or chemical weakness—micas, feldspars, pyroxenes, hornblende, olivine—are inexorably broken down, converted into tiny particles of clay and rust, and washed out into deeper water. Only the most resistant minerals—chief of which is the pure silica of quartz, that defies most chemical weathering—remain.

Sand, therefore, is mostly of grains of quartz; some particles are even more resistant: well-nigh indestructible zircons and monazites, rutiles and tourmalines. The pebbles in the gravel are also commonly rich in silica: fragments of sandstone and flint, and milky-white eroded fragments of vein quartz, concentrated a thousand-fold because fragments of weaker rocks have long been destroyed. These are the winners in the durability stakes, some of which have passed through eons-long cycles of erosion and abrasion in the growth and destruction of mountain chains—and survived.

Now, the particular beach upon which those grains are currently being washed and rolled by the waves (and upon which we, at play with volleyball or beach cricket, unconsciously create a whole variety of frivolichnia each year) is not going to be dug up to make the foundations of a multi-storey car park. Leisure being the biggest industry of contemporary advanced human civilization, the local Tourist Board and Hoteliers Association has more than enough muscle to see to that. But there are many examples of ancient beaches a million or so years old, dating back to some time when sea level was higher, and these are stranded high and dry upon our landscapes. As these are not (currently) in delectable seaside settings, it is easier to get planning permission to extract sand and gravel from them.

Beaches are extremely good sand and gravel factories, but they are not the only ones. The remains of ancient river-beds, and the meltwater plains that spread out in front of Ice Age glaciers, were equally efficient production lines. Perhaps more to the point, fossilized beaches, river-beds, and meltwater streams made of rock that was once sand and gravel can be found in rock strata dating back to, well, when recognizable sedimentary rock strata appear in the Earth's geological record, which is well over three thousand million years ago.

Thus, the main component of concrete, particles of sand and gravel, dominated by quartz, has inbuilt geological durability. It is likely to be a survivor within the underground realm, though, as we shall see, it might not survive *entirely* unchanged. But what about the other third?—the shale and limestone that is mixed in with the aggregate, to be then converted into cement.

These ingredients, of course, have equally good geological credentials, though the manner of their combination is novel, as these form various calcium silicates and aluminates in the firing process ('clinkering') and the hydrated equivalents of these in setting. Their long-term underground alteration, though, may be speculated on, and is of particular interest to those who plan the underground storage of nuclear waste. Tests here include 'accelerated weathering' in specially chosen chemical solutions, and some results indicate that, over millennia, the calcium hydroxide in the mix can be leached out by water, while acid groundwaters may form soluble calcium and aluminium chlorides. This would weaken the structure, producing an effect akin to osteoporosis in bones. Any fluorine, though, can lead to the formation of calcium fluoride, which is highly insoluble. Cement is certainly one of the more interesting experiments in long-term artificial rock formation.

A variety of other ingredients may be found in the Urban Strata. Bricks and tiles, for instance, are omnipresent pretty well anywhere where humans build habitations. At their simplest, these are simply oblong or flattish cakes of mud. But lithified muds—mudrocks—are the most common constituent of ancient sedimentary rock strata, and have little problem in coming to terms with the challenge of near-eternity. Mud is the end-product of the weathering of other kinds of rocks and is made of various proportions of a few ingredients: clay minerals; tiny fragments of resistant mineral such as quartz; some lime (optional); and variable amounts of organic matter upon which an interesting array of bacteria will be feasting for quite some time (indeed, the toughest of the bugs will still be hanging on to life at burial depths of a kilometre or more).

The mud in industrially made bricks, though, is a little different. It has been rapidly heated in kilns so that the minerals alter and partly fuse to produce a usefully hard and resistant new material. A natural analogue, of a sort, is mudstone which is baked by coming into contact with ascending magma, such as might happen beneath a volcano or adjacent to an underground magma chamber. It metamorphoses, new minerals growing from the original tiny flakes of clay mineral. Indeed, bricks, in many respects, might, at first glance, seem to be more resistant to being changed during burial than the original mud. For instance, mudrocks that have been naturally baked by underground magma have increased resistance, say, to the massive pressures inflicted when continents collide and throw up mountain ranges. Thus, it is the contact metamorphosed rocks that are more likely to preserve vestiges of their original structure, while their unbaked neighbours are sheared and stretched by the immense pressures, so that little of the texture of the original mud may remain.

The natural high-temperature baking of mudrocks typically takes centuries to many millennia to complete. A brick, though, is only fired in a kiln for a day or so, albeit to temperatures higher than that of a granite magma. This flash-heating dehydrates the original clay minerals, to produce anhydrous high-temperature minerals such as mullite, cristobalite, olivine. The new minerals that grow form tiny crystals, held together by a glass formed by the melting of a few per cent of the original raw material. Newly formed brick is also porous, being traversed by a network of pores and cracks that have formed, allowing vapour from the dehydrating brick to escape during the firing. Related to bricks are ceramics, that are also fired clays, though of greater purity and consistency. As in bricks, the brief, intense heating dehydrates many of the original minerals.

Glass is another common ingredient of our material civilization. It is a frozen liquid, that starts as a silica-rich melt which is then cooled by water jets so quickly that the constituent molecules do not get the chance to lock together, Lego-style, into crystals. Humans make glasses in glass factories, from pure silica sand, together with sodium carbonate as a flux to make the silica melt at lower temperatures, and calcium carbonate, to keep the resultant window glass from dissolving in the rain, for sodium silicate on its own ('water glass') is water-soluble.

There are quite natural glasses, such as obsidian. This also starts off as a melt, one so silica-rich that it is highly viscous (the silica molecules tending to polymerize, even in the molten state). These lavas move only slowly as a stiff, treacle-like liquid; at a microscopic level, the molecules also move only slowly and with difficulty. In any situation experiencing rapid natural heat loss—as in small lava flows or thin sheets of magma intruded into cold rock—crystals, again, do not get a chance to form.

Over geological timescales measured in tens of millions of years, such natural glasses are temporary. The individual molecules are more stable when arranged as into crystals, so even in the solid state they migrate, infinitely slowly, the necessary tiny distances to lock them together with other molecules to form microscopic crystals. As these form, and crystal faces begin to scatter light, the obsidian loses its glassy character. It devitrifies, changing into felsite, a paler coloured, opaque rock.

It is likely that artificial glasses will, over time, likewise transform and lose their transparency. Jam-jars and milk bottles and window shards, one hundred million years from now, will almost all be milky-white, opaque. Fragments of mirrors, unearthed by future excavators, will neither reflect the new reality of the future, nor their original function in the human present.

Then there are the metals of our empire: iron and steel, aluminium, copper, tin, zinc, lead, not to mention silver, gold, and platinum. Now these are a little more unusual, for there are not that many native metals in nature. There is gold, of course, prized for its lack of reactivity and hence its enduring nature. Copper can also be found native, but most commonly it and other metals are present in nature as compounds: carbonates, sulphides, sulphates, silicates, oxides, and hydroxides. Iron is rarely found in its native form—mainly as iron meteorites. So buried iron and steel objects, unlike concrete, do not mirror nature, or at least do not mirror the natural environment on and within the crust of this planet. Their longevity cannot be taken for granted. The underground realm may not treat them kindly.

Then, there are plastics, various polymerized hydrocarbons—nylon, acrylic, polystyrene, polyethylene. Fabricated as they are, these have eminently ancient precursors. Their component molecules were once part of the complex biochemistry of planktonic organisms that lived in now-vanished oceans. These died, fell to the sea floor, were buried in sediment. Then, as the sediment layers were transformed into hard rock strata, the organic matter broke down. Part transformed into petroleum gas and oil and migrated towards the surface, some to be trapped en route in hydrocarbon reservoirs underground, while part formed a resistant graphite residue left behind in the strata in which the original organisms were buried. Some of the oil, subsequently pumped out of the ground by human industry, was then refashioned into the plastics that help define our age.

What might be the best fossil analogues for, say, a polystyrene cup? Perhaps some of the long-chain organic molecules in the organic skeletons of some marine invertebrates. Fossil graptolites, for instance, might provide a useful analogy. These were extinct animals that formed colonies that made up part of the planktonic population of the high seas, some 400 million years ago. Geometrically baroque and biologically enigmatic, there is nothing quite like them in today's plankton. Of the animals themselves, little trace has ever been found. Their living quarters, though, are delicate matchstick-sized tubes made of a collagen-like substance (such as your fingernails are made of); they are among the most common fossils in marine rocks of that age.

The human empire has long, also, used natural products: never more, indeed, in absolute terms, than now. Wood and to some extent its derivative, paper, and also textiles, have well-known natural analogues. Fossil wood is not uncommon, while even leaves and twigs may be found impressed into strata, and in bulk, of course, it forms coal. There are differences, that may prove significant. Human-worked wood is typically dried, seasoned, often varnished or lacquered—whether for a kitchen table or a concert violin. From the start, therefore, it is more decay-resistant. It is a helpful first step on the long road to eternity.

So these are the ingredients. Knowing their pedigree, one can guess their future. It is time to bury them, and see what happens.

SHALLOW GRAVE

The ingredients of our cities are unlikely to survive unchanged, for the most part. The changes that ensue will produce humanity's very own addition to the standard geological catalogue of rocks and minerals. That is, given a little time (the chosen, but entirely arbitrary hundred million years), a little heat (between, say one hundred and several hundred degrees centigrade, depending on how deeply they are buried and how hot the Earth's local crust is), some pressure (the weight of overlying sediment, up to several kilometres thick) and the catalytic, corrosive effect of the warm and then, eventually, scaldingly hot, chemically charged underground fluids in which all of these structures are to be bathed.

The starting point? One might take as example the burial of New Orleans—or of Amsterdam, Haiphong, Venice. Indeed, any one of the myriad coastal plain cities could serve equally well. Burial will be untidy. It will likely, from today's perspective, begin via a sea level rise of somewhere between two and five metres, in the next century or so. Geologically trivial, this will be nevertheless enough for widespread inundation and abandonment of substantial areas of urban landscape.

The substructures, early on at least, are likely to be in relatively good shape. Their upper structures would almost certainly rapidly become smashed, collapsed and corroded, as distressing a sight as, say, a dinosaur lying rotting in a Jurassic swamp. But, no matter: the bones of that dinosaur, later fossilized, might have gone on to grace the Carnegie Museum. Likewise the skyscrapers, from street level down, would stay in something like their original shape, amid the rubble collapsed around them, as the curious fish glide in and out through the broken windows. Their basements and foundations would remain more or less intact, as the mud drifted over them.

The deep skyscraper roots form inverted concrete and steel spires beneath a New Orleans that is slowly sinking into the Gulf of Mexico, as the detritus of half a continent, washed on to the top of the Mississippi delta, presses down on the malleable crust. Around the tops of the concrete piles snake the thick tangle of pipes for water, electricity, gas, sewage, optical cables, of subways, underground car parks, and nuclear fall-out shelters. Once in the burial realm, these abandoned foundations of our human empire can begin their transformation into the Urban Stratum that may, in the yet more distant future, be discovered, analysed, explored, marvelled at.

A millennium on, these foundations may lie in water that is twenty metres deep or more, as the ice caps continue inexorably to melt. Overlain and surrounded by storm-strewn rubble, they may now also lie under a few metres thickness of sand and mud, washed in by waves and currents. This will now resemble an archaeological site, akin to that of some ancient Greek or Egyptian city drowned by tectonic subsidence following an earthquake or volcanic eruption. Changes will begin. Anything near the sea floor will be within reach of burrowing worms and sea urchins and crustaceans. These will not only begin to rework anything nutritious, such as paper and textiles, but will also, through their burrows, allow access for oxygen-bearing seawater and oxygen-respiring bacteria. Anything made of wood will now be waterlogged. If it is buried deep enough to restrict the access to oxygen, then decay may slow substantially and a good deal of the wood's original strength may be preserved. In areas of the English Fenland, bog oaks, buried in peat layers for thousands of years, were highly sought-after building materials in historical times. This is, indeed, the start of the formation of coal.

Brick and concrete will also be waterlogged. Concrete may not appear much changed, but the bricks will likely be beginning to alter, very subtly. They will be getting bigger. Firing a brick drives off the water. Once in a wall and exposed to the elements the brick will absorb water, and expand: only by a fraction of a per cent, but, over time, sometimes enough to fracture it, especially if hard cement-based mortar is used in construction (rather than softer lime-based mortar that can better absorb the expansive stresses). This expansion can carry on for thousands of years (it has been observed for Roman bricks) albeit at progressively slower rates. Nevertheless, once in the burial realm and permanently waterlogged, the bricks of the Urban Stratum will expand to an effective maximum, and the stresses involved may break many of them.

Underground chemical changes will begin, too. Metal objects will be subject to these. Take iron, for instance. In moist air it rusts, so burying iron or steel objects below the oxygen-rich surface might be expected to protect them. It's not necessarily so. Buried archaeological artefacts made of iron are often pitted and partly dissolved. Iron transforms, and one of the iron-related reactions produces a bright golden lustre—one that is misleading, though, for any prospector in search of an earthly fortune. Just below the sea floor, where the oxygenated environment of the surface passes down into the anoxic, reducing conditions of the subsurface, iron and sulphur are on the move. The iron is that which has become reduced to its mobile, ferrous state. The sulphur is that which originally formed sulphate ($SO4^{2-}$), one of the major salts dissolved in the oceans, but which was stripped of its oxygen atoms by energy-hungry bacteria, just below the redox horizon in the sediment, to form sulphide (S^{2-}). The ferrous and the sulphide ions link up to form iron sulphide as the mineral pyrite, or to give it its vernacular alias, fool's gold. Layers and crystals of this are common in sea-floor sands and muds. Once formed, they persist within the strata as long as conditions stay reducing.

Pyrite tends to form in subsurface cavities. In the chambers of ammonites, or within the fragile empty skeletons of the entombed graptolites, often filling the entire space to create perfect replicas of their interiors, three-dimensional metal sulphide casts of remarkable strength and durability. A pyritized graptolite can survive even within mudrock that has been caught up in mountain-building events: the original mudrock has been sheared, recrystallized, transformed into slate. But the encased pyritic graptolites, rigid and resistant bodies within the plastically deforming rock mass, can retain even microscopic details of their original structure. They may break at weak points, be pulled apart as the mountains are raised, yet still the disarticulated fragments preserve the original shape of the long-dead animals' living chambers.

Only when the fossils are carried back towards the oxygenated surface does this stability disappear. In landscapes made of graptolite-bearing Welsh slates or ammonite-rich Jurassic shales, surface water seeping into the rocks begins to attack the pyrite, oxidizing it to friable or powdery iron hydroxides, that may then simply disappear, washed away like dust. The three-dimensional cavities remain in the rock, though, and preserve, just as faithfully, the shape of the original fossil. A palaeontologist, filling the empty cavity with liquid latex and allowing this to harden into a flexible rubbery compound, and then gently pulling out the cured latex, can re-create a replica of the solid interior cast of the fossil, that had so long been guarded by the pyrite in the underground realm.

Wherever cities are buried in mud, pyritization will take place. What might the fossil-equivalents be? Small part-empty plastic containers, perhaps, and sections of plastic tubing and wire. The surfaces of fragments of cup and flowerpot. The interiors of any of the myriads of tiny metal and electronic gadgets that we now produce in their millions also seem to be good candidates for such sulphidic coating and infill, for these in themselves contain iron, one of the ingredients of pyrite. Part of the detritus of human civilization will certainly bear the sheen of fool's gold.

Some of the metals of human manufacture, though, are carefully designed to resist chemical attack. The steel that is laced with chrome or molybdenum or vanadium to make machinery to survive the attack of the elements around oil rigs or on battlefields, or to make stainless steel kettles and cutlery. Many of these might survive the early stages of burial, at least sufficiently to make an impression on the compacting and hardening sediment around them; it is an intriguing question, then, as to exactly how they would survive longer-term burial.

It is a similar case with aluminium, an abundant terrestrial element (making up some 8 per cent of the Earth's crust) that is never found native, always being chemically combined with other elements. It took people a long time to discover how to purify it; but, once smelted, it forms a strong, lightweight metal that, somewhat counter-intuitively, is corrosion-resistant. This is because of an oxide film, just a few atoms thick, that forms on the surface and prevents further oxidation. It is highly effective protection: scratch the film down to bare metal and it almost instantly re-heals. Titanium behaves similarly, and is even more durable in normal surface conditions. Interestingly, these oxide films degrade in oxygen-free conditions, allowing the metal to begin to pit and corrode. Again, one is curious to know just how aluminium and titanium objects would survive in the reducing conditions of the burial realm over geological timescales. Of other metals, copper and zinc are more soluble and mobile than, say, lead. In fossilization, therefore, the electric wiring could well disappear while the plumbing remains.

DEEP BURIAL

Fast forward a few million years. Our coastal cities now lie deeply buried. Entombed beneath layers of mud and sand that may in places be hundreds of metres thick, the masonry and its contents begin to be squeezed, flattened, distorted by the weight of the billions of tons of sedimentary layers accumulating above them. Physically, the structures of the Urban Stratum are now being crushed to varying degrees. Those buried in mud will be crushed most, for this substance is loose and fluffy to begin with, comprising mostly water. It then loses up to 90 per cent of its original volume as it progressively compacts down on being buried ever more deeply. Such physical compaction is less for those objects buried in sand, which initially only loses around a quarter of its volume as the grains readjust their positions to pack a little closer together.

Compaction alone, though, should not altogether obscure the identity of the objects of human manufacture. After all, delicate ammonite shells and trilobite carapaces, crushed flat within ancient mudrocks, are still perfectly recognizable to fossil collectors, even if they are broken into many pieces. This simply gives more of a jigsaw puzzle to whoever may study our remains in the far distant future. The challenge is not always as great as it sounds, for the jigsaw usually comes ready-assembled, as the component fragments of the fossil, encased within the rock, generally remain more or less in their original relative position.

There will be sheltered zones, too, here and there protecting artefacts from the pressure, much as the hull of bathyscaphe today protects the fragile bodies of submariners from the crushing pressures to be found on the ocean floor. In a buried city the protection is afforded by anything that still can retain its strength and rigidity: the concrete walls of a well-built cellar, or perhaps of a nuclear fall-out shelter, may resist the pressures imposed by the ever-thicker and heavier drape of sedimentary strata. The surrounding highly pressurized mud layers will deform around such constructions, slowly flow around them, may even be injected into them through fractures and crevices to eventually fill the spaces that were once just air-filled and then became flooded with water. Yet still, any objects within those spaces—abandoned furniture, cups and saucers, machinery of whatever sort—will be spared from crushing, and will—at least physically—retain their three-dimensional shapes.

Such differences in compaction are commonplace in buried strata, of whatever age. Individual buried ripples or dunes of sand, encased within mud, will retain much of their shape as the compressible mudrock slowly deforms around them. The fragile shells of fossil ammonites, encased in isolated, early formed carbonate concretions—a kind of natural concrete, as it were—can be hammered out of those concretions in their full three-dimensional splendour, while the surrounding ammonites, lacking such armour, are flattened on to the single plane of a lamina of mudrock.

There will be chemical changes of the buried artefacts, too, though these are more complex and difficult to predict than the physical changes. The subterranean realm is permeated by water, that most efficient of solvents. Much of that water started out as primeval seawater, carried down in the spaces between the original sedimentary particles. This buried water is then forced back towards the surface, as those sand grains and mud flakes are squeezed closer and closer together by the pressure of the overlying strata. Much of the fluid expulsion, and consequent volume loss, takes place over the initial stages of burial, say to a few hundred metres depth below the surface, and after that ever more slowly. The water, forced along slow, tortuous paths through the ever-narrowing spaces between those grains, is continually dissolving mineral matter as it migrates through the rock. This mineral matter then often crystallizes out later along that subterranean path, between other sediment grains, thus cementing them together. That is why some fossil shells, originally calcium carbonate, are replaced by silica, or simply left as empty (but still recognizable) spaces within the rock, and why loose sediment, once underground, is changed into solid rock strata.

The slowly migrating water varies in chemistry, which is determined by the nature of the strata that it is being filtered through. Where it is acid (say, because it has been filtered through a mud in which organic material is breaking down) it may begin to dissolve the calcium carbonate cement within concrete and mortar, turning some of these into loose masses of sand grains (that themselves, though, largely resist such chemical attack). These masses will be held in place by the stratal layers above and below but will fall apart once excavated. They will be like many dinosaur bones: once strong enough to allow the living saurians to be ferocious and athletic predators, and now more friable than biscuit, needing careful encasement in burlap and plaster of Paris to allow their removal from the rock and then their transport to a museum.

Buried bricks within rubble and the remains of walls and buildings will likely, also, change in texture. A brick buried in mudrock may, for a while, retain something like its original hardness and shape (albeit slightly expanded through hydration), the soft mud deforming around it as the stratum is compacted. As this happens, the brick will become more porous than the compacted mudrock around it, and so will tend to act as a natural conduit for those slowly but constantly migrating underground fluids. The human-fabricated minerals will be unstable in a fluid-soaked mud blanket that may be now as hot as a just-poured cup of tea, but will still be over a thousand degrees centigrade cooler than the temperature of the furnace where the bricks were fired. The minerals will slowly break down, back, probably, into something like the original clays, while new minerals may well grow in the original pore spaces of the brick, rather like tiny stalactites slowly filling a network of microscopic caves. A fossil brick—and ceramics likewise—could, millions of years hence, alter to a substance softer and crumblier than the mudrock around it. It may then flatten somewhat if the surrounding mudrock continues to deform, and may change colour too. The red colour of many bricks is due to oxidation during firing; on long-term burial this should reverse, with fossil bricks reverting to the greys and blues of the original raw material. With ever-deeper burial, the tiny, highly reactive flakes of clay mineral within the original bricks will change their structure, recrystallize, eventually transforming into larger (but still microscopic) crystals of different clay minerals; as yet higher temperatures are attained, these clay crystals begin to transform into mica. Yet—and here is the bottom line—these brick fossils, and the concrete fragments lying next to them, should still be recognizable as artefacts.

And then there are those plastics that resisted the earlier biodegradation. What might happen here? Let us assume that at least some of these are akin to the complex, resistant hydrocarbons that, say, form the tough outer coats of pollen and spores and the skeleton-homes of the graptolites. These gradually transform in the natural pressure cooker of the deep burial realm. The molecules break down, losing the volatile components—chiefly hydrogen—as methane and other light hydrocarbons that make up natural gas. As this happens, a pollen grain, on being ever more deeply buried, changes colour from its original pale translucency, through progressively darker shades of straw-yellow to orange to brown, eventually turning to the opaque black of the graphitized carbon shell it finally becomes.

This colour change is so predictable that it has become, for modern palaeontologists and petroleum geologists, a kind of palaeo-thermometer for strata, that tells them how much burial and how much heating they have experienced. At least some of the human-manufactured plastic artefacts, in the millions of years to come, should undergo a similar colour transformation, and strata-encased plastic cups and shampoo bottles will have forever lost their pristine transparency. It is a useful colour series, this, one of the chief clues to another transformation: the underground creation of oil.

At depths of about two kilometres, perhaps tens of millions of years after burial, large complex molecules of entombed organic substances are broken down into smaller molecules, of just the right size to form tiny drops of petroleum oil; these migrate upwards, either to reach the surface, or to be trapped underground in hydrocarbon reservoirs. The pollen-colour geothermometer tells petroleum geologists whether or not the rocks have been cooked up enough to yield oil—or cooked up so much that the oil itself has been broken down.

For some part of the plastic detritus of human civilization, this will be the ultimate form of recycling. Whether or not they leave behind them the graphitized carbon ghosts of their original form, much of the original material will be broken down and merge with the much greater mass of hydrocarbons derived from the organic detritus—the remains of algae and bacteria—in the mudrocks. One may predict a range of fossilized plastic tubes and bottles among the Urban Strata, varying in colour from pale yellow to brown to black, depending on how deeply they were buried. Other plastics might, under heat and pressure, break down completely into their component molecules, which migrate up through the rock strata until they are trapped by some impermeable rock layer, to become a minute fraction of newly forming oil and gas reserves. These hydrocarbons thus formed could, conceivably, help power the far-travelled civilization that might one day excavate the remains of our cities (though our future explorers will likely have more effective energy sources, and may be less cavalier as regards perturbing the carbon balance of an entire planet).

FURTHER TRAJECTORIES

Over a hundred million years underground, much can happen. There are many possible trajectories and scenarios. Some strata may be slowly buried to only a few hundred metres depth, and then as slowly be lifted back to the surface, never even reaching the temperatures and pressures that allow the formation of oil. This is effectively what has happened to the Oxford Clay, of Jurassic age, of the English Midlands (this is not much more than a compressed mud still, scarcely deserving the term of mudrock, and one can peel the layers apart with one's fingers to expose the wealth of fossils it contains).

Other units of strata may be buried to several kilometres depth, and transform into respectable rock. Yet others may, like the strata of the Alps and the Himalayas and the Andean cordillera (much of which are considerably less than a hundred million years old), be caught up in mountain-forming regions. The rocks would be crumpled, sheared, more thoroughly recrystallized. In the outer parts of mountain belts, where mudrocks become slates, and sandstones become quartzites, sedimentary structures and fossils—and by extension some echo of the structures that might comprise an Urban Stratum—are still recognizable. Once in the central and deepest parts of a mountain belt, though, where temperatures rise to several hundreds of degrees at depths of tens of kilometres, the rocks thoroughly recrystallize, to become schists and gneisses. Ultimately, part of the rock begins to melt, producing migmatites, and then the melt separates and accumulates to form granite bodies that slowly migrate back towards the surface, shouldering aside the more solid rocks that lie in their path. Any city that finds itself in such a region, the seventh circle of Hades, is gone forever, transformed into wholly recrystallized lenses of melt-derived rock and isotopic wisps. At this stage of recycling, its original identity disappears.

Some cities will experience such a destiny, but they should be in the minority, for at least the next billion years or so. Many fossilized cities will remain within sedimentary strata, and will stay within the sedimentary realm. They may be tilted or gently crumpled by earth movements, or dislocated along fault lines. No matter; such things are workaday complications for any geologist, whether Earth-born or not. Occasionally, petrified cities will be exhumed back to the surface. Here, there will be metres-thick layers of rubble, of compressed outlines of concrete buildings, some still cemented hard, some now decalcified and crumbly; of softened brick structures; of irregular patches of iron oxides and sulphides representing former iron artefacts, from automobiles to AK-47s; of darkened and opaque remnants of plastics; of white, devitrified fragments of glass jars and bottles; of carbonized structures of shaped wood; of outlines of tunnels and pipes and road foundations; of giant middens of rubble and waste. Humans have introduced a fascinating and wholly new set of materials and structures for fossilization processes to act on. It is certain that a good deal of such things can, after burial, survive in some fashion almost forever.

When that encounter happens, our future explorers will have a bonanza to explore, to interpret and misinterpret, a palaeo-archaeological goldmine into deep history and long-vanished intelligence. What are the chances, though, of their encountering the mortal remains of the builders of this

distant empire, and of understanding their actions and motives? Such encounters might well occur. Then, our explorers will be able to compare the scale of the environmental and cultural phenomena that we have created with the form of our own modest frames. There may be, in the future, a good deal of perplexity, and some quite outrageously misleading deductions, about the nature and character of human beings.

It is time to prepare for the final meeting.

Body of Evidence

A breakthrough! We have, at last, identified the city-builder. A small to medium-sized animal—one species only, our biologists think, but puzzlingly distinct from others of the time. Large-headed for its bulk, seemingly upright in posture, obviously manipulative. We presume local to the planet, but even this is disputed. We understand the associated stratal layer a little better now, and can better predict its course; other skeletons are being found. There is excitement, of course, but these remains are generating, also, more complex responses from our scientists. The excavation in some way is a kind of meeting, too: one for which we have arrived too late. Nevertheless, we wish—we must—find out more about these beings. Our analysis proceeds.

THE BODY IN FOCUS

The most direct legacy that we can leave to future geology is that of our own mortal remains. Today, in reconstructing the long-vanished Jurassic landscapes, we put the mighty, charismatic dinosaurs square in the foreground. This focus we have—well nigh a fixation—seems to us almost self-evident. Were they not the rulers of their empire, just as we are of ours, literally bestriding their domain as colossi of scale and blood and bone? Their skeletons, avidly sought, intensely studied, painstakingly reconstructed in museum displays, are the symbols of those times, iconic, mesmerizing. Might we not hope for similar awe and reverence from our future excavators?

There is no guarantee, of course, that these as yet unborn explorers of a future Earth will share this perspective. Perhaps their focus will be on what, among all the diverse living inhabitants of this planet, is most important in preserving this living tapestry. They may well regard the myriad tiny invertebrates, or the bacteria, of the world as much more important to that (in planetary terms) rare phenomenon, a stable, functional, complex ecosystem.

If these future explorers took this view, at the risk of offending what little there might then remain of our *amour propre*, they would have a point. Take away the top predator dinosaurs, and the Jurassic ecosystems would have been a little different, to be sure, but no less functional. Take away humans, and the present world will also function quite happily, as it did two hundred thousand years ago, before our species appeared. Take away worms and insects, and things would start seriously to fall apart. Take away bacteria and their yet more ancient cousins, the archaea, and the viruses too, and the world would die.

But, let us imagine our excavators as being, in true science fiction style, just as obsessed with their relative position in the food chain as we are. Let us assume that, in their excavation of the Earth's history, they will be looking for the power brokers of the ancient past, that they will be digging for bones and bodies. What chance, then, human remains?

It is not safe simply to assume, because we are familiar with finding fossils from hundreds of millions ago, that we as a species will leave a viable long-term fossil record. It is a little sobering that, despite the many active professional palaeontologists of the last century and plenty of interested amateurs, we have found representatives of only a tiny fraction (some 0.01 per cent, at a rough but reasonable estimate) of all species that have ever existed. Statistically, then, the odds are stacked against us. Fossilization is a game of chance, and the odds are not good. But fossilization is poker, not roulette. There are things that a species can 'do' to improve its chances of immortality. So, let us examine whether or not the deck is stacked in our favour.

We will assume—perhaps to the chagrin of those individuals who have expensively arranged to be deep-frozen in hope of future revival and cure—that our remains will be thoroughly, absolutely, emphatically stone-dead. The immortality we are discussing is of a kind that is, alas, post-mortem. For those lower in the food chain, that rule may, astonishingly, not necessarily apply. It has been claimed, in reputable scientific journals, that bacteria can survive in suspended animation, within tiny bubbles of water in underground salt layers, for as long as a quarter of a *billion* years. This is controversial, and may yet be disproved; but it remains a possibility. We, alas, are not bacteria. So our journey must perforce start with death. What then, is the norm, after death, for a terrestrial multicellular organism?

Soft tissues, almost always, disappear quickly after death, as necrolysis, the organic decay of a carcass, takes place. The process starts without any outside help. The enzymes in organic tissue, which, in life, carried out its essential chemical reactions, will turn on their own cells and begin to use them as a source of fuel. This process, called autolysis, will cause degradation of a carcass, even if the carcass can be protected from all the various kinds of living organisms which see it as a source of energy. Bacteria and fungi, in particular, those great, indispensable recyclers of all of the chemical elements that make up living beings, will rapidly putrefy and decompose a carcass. The process will be helped along by animals, such as the worms of folklore, which will also see the tissues as a source of food and will scavenge what they can. In environments where necrolysizing organisms thrive, bodies can be reduced to skeletons extremely quickly. Indeed, at times, astonishingly quickly.

Take the graptolites of some 400 million years ago, those colonial, matchstick-sized zooplankton, roaming the Ordovician and Silurian oceans. These are known from their hard proteinaceous skeletons, of which literally millions have been collected and examined; the soft parts have almost never been found. Why not? An experiment was performed a few years ago to find out. While there is nothing like the graptolites in today's plankton, there are some living relatives, the pterobranchs, that live, here and there, on the sea floor, and that also have soft bodies and tough tubular skeletons. Just-dead pterobranchs were, quite simply, placed in a tank of seawater to see what would happen. The wait was not long. After just a few days, the soft bodies had disappeared completely, fully decomposed. But then the observations became a little more tedious. The skeletal tubes persisted. And persisted. And persisted. Three months into the experiment, they were still there.

It just goes to show the value of a good skeleton. Consider another example: the sea cucumber, a relative of starfish and sea urchins. This is an obscure creature, not least to most palaeontologists, though not because it is rare. It lives on the sea floor, and its numbers far exceed that of the human population. Sea cucumbers live in shallow seas, which are in general a good place to live to increase the chances of being fossilized. They are also known to live in huge numbers in deep water, so their distribution is arguably even broader than ours, as the ocean covers a much greater proportion of the Earth than does the land. Despite this, the fossil record of sea cucumbers is, for all practical purposes, nil; a couple of likely examples accompany the *Archaeopteryx* in the Solnhofen Limestone, but that is about it.

Most modern sea cucumbers are entirely soft-bodied. They have no bones or shell, which helps explain why they are considered a delicacy in some countries. The skin of a few modern species, however, contains tiny, hard plates made of calcite—crystalline calcium carbonate—making them less appetizing but eminently more fossilizable. Indeed, the fossil record of sea cucumbers consists almost entirely of these minuscule plates.

The reason is not hard to find. The soft parts of sea cucumbers have been food for hundreds of millions of years before the sushi chefs discovered them. The oceanic predators and scavengers that fed on them, as well as the bacteria involved in the decay process, however, found their small calcite plates indigestible, and so these could be left for posterity.

Yet, predation, scavenging, and decay are not the only challenges facing any sea cucumbers with aspirations of immortality. There is a whole suite of processes that can affect them between the moment of death and their burial in layers of sediment, the study of which is called taphonomy. Taphonomy translates as the 'laws of burial', though the 'laws' here act as guidelines rather than rules strictly adhered to. There are instances in which, unlike in the ironclad laws of thermodynamics or planetary motion, nature decides that the normal conditions for preservation need not apply. A first taphonomic commandment might be something like: to enter the fossil record, hard parts are needed, such as bones, shells, or small calcite plates. The sea cucumbers may have held closely to this rule, but there are exceptions, as we will see.

Between the sea cucumbers and humans, there is almost a world of difference. This difference, to be precise, is the surface of the sea: the sea cucumbers inhabit the marine realm, while we are land-based vertebrates. The rules, therefore, are different. We need to examine what happens to a terrestrial vertebrate in nature, and then move on to considering quite how natural we are, when it comes to disposing of the bodies.

There have been, a little ghoulishly, quite a few studies of what happens to dead bodies of vertebrates in the wild, whether they are sheep on Welsh hillsides, or antelopes and wildebeest in the plains of Africa. Let us run through taphonomic history of, say, an average wildebeest. At the terminal stage in its life it will likely become, rather abruptly, a meal. The pride of lions that feeds off an unwary (or slow, or old) wildebeest will eat its flesh, but not, for the most part, consume its bones, though these are commonly scraped, scratched, or broken in the feeding frenzy. Then, the vultures and hyenas that come to scavenge the remains often transport parts of the body or individual bones. Porcupines are bone thieves, too. So, the skeleton could well be in fairly poor shape even before it is subjected to the ravages of the weather. Not all wildebeest, of course, are predated. Some may die of disease (though they will then be likely scavenged) or be swept away in a flash flood (to become the focus of attention of crocodiles). In one classic taphonomic study of wildebeest, it was found that, given a starting population of 1 000 wildebeest, only 50 individuals would be represented in a buried assemblage and, of the 152 bones that composed the wildebeest's skeleton, on average only 8 bones per carcass would be found.

Imagine, therefore, that you have never seen a wildebeest and now you have to reconstruct what a herd looks like from 5 per cent of the population and 5 per cent of their bones. One can forgive palaeontologists for becoming a little frustrated at the amount of information lost between the time of death and that of burial (and this is not counting the further information lost between the time of burial and the time that someone actually gets around to study those wildebeest fossils).

Yet there is consolation in that, while information about the wildebeest population has been lost, information has been gained about the environment in which they lived, died, and were buried. One might, for instance, be able to infer the presence of the substantial predators—and their jaw dimensions—from teeth marks on the bones, or details of the climate from their weathered condition.

Wildebeest are common herbivores that in their natural state form herds tens of thousands strong. Their equivalent among extinct mammals might be the aurochs and Irish Elk. In the era of dinosaurs, comparable animals might have been iguanodons and stegosaurs. These are common and familiar as dinosaurs go—but then one has to recall that dinosaur skeletons have been sought with great determination and effort over some two centuries by many hundreds of palaeontologists, professional and amateur, from over a hundred million years' worth of strata. Yet, skeletons that are anything like complete amount to only a few thousand specimens in total.

PRIMATE POSSIBILITIES

How do humans compare with wildebeest? One may take first the perspective of a modestly deep span of time, a few million years or so. In this interval of the recent past the studies of palaeontology and palaeoanthropology combine, and different primate species that are potential human ancestors are sought, quite as avidly as the dinosaurs are hunted in older strata, for we are intensely curious about our origins.

Here, the pickings are thin. Palaeoanthropologists are apt to break into a bottle of champagne when a single bone fragment is found, while very early hominid finds become major news stories. Lucy is so recognizable because there just are not that many other early human ancestors to confuse her with. The hobbit, with one discovery and one well-crafted simile, became as well known with respect to human ancestry as for Tolkien's fantasy, when the new species *Homo floresiensis* made tabloid headlines around the world. And it was only while these pages were being drafted that the first fossil chimpanzee was discovered.

There is worse to come, from the point of view of geological longevity. These finds, seldom as they are, were made in strata that are geologically young and therefore near the surface and accessible, and hence also widespread. They are also generally in terrestrial settings, and so in the realm of long-term erosion. That most famous cradle of pre-human fossils, the Olduvai Gorge, exposes strata deposited in a medium-sized lake just before the Ice Ages, when sub-Saharan Africa was wetter. The active erosion helps modern humans find their ancestors today but, geologically soon, erosion will likely remove most or all of these deposits. Resting on the present-day African landscape at an altitude of some two kilometres, these strata are themselves firmly in the erosional realm, and thus effectively in transit. Geologically speaking, they are just passing through.

Homo floresiensis was found in a cave on the island of Indonesia, a cave that is also temporary when set against such a timescale. Over timescales of tens of millions of years, that Indonesian landscape will be eroded, and the caves in those hillsides will disappear also. So, one hundred million years from now, there will be much less to go at, and any search for direct human origins would almost certainly be fruitless.

But the search for historical and modern humans? Here the odds may be stacked differently. We may start by looking at the record through the eyes of an archaeologist. It is a useful start, even though the relics available to this type of study have only gone through a few of the barriers that need to be passed through to obtain truly long-term fossilization.

First, there are our sheer numbers, now. The explosive growth and geographical expansion of the human population is a biological phenomenon without precedent. The species *Homo sapiens* is about 160 000 years old. For most of that time, it was confined to Africa, and numbers were probably counted in a few millions. Some 80 000 years ago, modern humans spread to Asia and Europe. Forty thousand years ago humans reached Australia and then some 12 000 years ago the Americas, as the ice sheets of the last glaciation receded. By the time of the Romans, there may have been some two hundred million humans on the planet. Numbers reached a billion about two hundred years ago, then two billion by about 1930, and then three billion by 1961. Now it is some six and a half billion. By the end of the century human numbers might double again, though most estimates merely add another three billion or so.

This population growth has unfortunate social consequences, including famine, war, environmental destruction, and, down the line, a likely mass extinction in which we ourselves may well figure. It is good news, though, for species fossilization potential. In the same way that joining a syndicate and buying more tickets increases your odds of winning the lottery, the more of us there are, the better the chance that some of our remains will be preserved.

On the other hand, humans are animals that live on a land surface that is mostly being eroded, and not on a sea floor where sediment mostly accumulates. But, we, as a species, have done our best to compensate for this by living in habitats that are as diverse as possible. Some of us have picked better places to live, in terms of fossilization potential, than others. In the game of mountain goat versus hippopotamus, the hippo is going to win, simply because the mountain goat lives in an area of active erosion and the hippo lives in one where there is ongoing deposition. The human species, ecologically, includes both mountain goats and hippopotami, a quite remarkable feat of adaptability. So, to win this particular lottery, do not join the Swiss or Nepalese fossilization syndicates; fall in with the Dutch or Côte d'Ivoire ones instead.

Consider the past few thousand years of the archaeological record, and we might well get the impression that humans would have a very good fossil record. There are places where skeletons abound. The killing fields of twentieth-century wars, for instance, and the medieval cemeteries chock-full of plague victims; the Roman catacombs, and the burial mounds that dot the landscapes of Europe and North America.

Here, along with our increased numbers, there is a cultural practice that, from the viewpoint of an alien explorer might seem a little, well, *odd*, which is likely to help our long-term preservation: that of burying our dead. Preservation is greatly enhanced by rapid burial. Placing the bodies of our deceased in a wooden box and under six feet of soil protects them from the scavengers and from the weather, advantages (if that is the right word) not available to the average wildebeest. Cultural practices vary, though. The ancient Egyptian practice of mummification (of at least the higher class members of that society) was a good short-term measure for preservation; medium term, it has run the gauntlet of generations of grave robbers, while long term it may offer little help. The Zoroastrians of India, whose bodies, after death, were traditionally left on *dakhma*, or Towers of Silence, to be picked clean by the vultures, have also changed the taphonomic odds. In recent years, though, there has been a population crash of vultures in Asia, caused by widespread use of the toxic cattle drug Diclofenac (now banned, but too late for the vultures), that has made this swift and eco-friendly cultural custom largely impossible. Nevertheless, in general, widespread human burial cannot help but increase human fossilization potential.

The link between humans and the Urban Stratum will not be *quite* straightforward. The taphonomy of humans here will be quite distinctive, and be utterly unlike, say, the Ice Age caves that acted as home and grave, sequentially, for generations of bears and wolves. Modern cities, at least those that more or less function as human habitats, contain few human bodies, other than a small number, perhaps, of murder victims buried at midnight in back gardens; it would be extraordinary luck for a future palaeontologist to find one of these. Indeed, if a city is buried in the state it was in as a functioning life assemblage, and abandoned in an orderly fashion as sea level rises, then those living quarters that will fossilize will be remarkably free of human skeletons, and resemble nothing so much as a vast geological Mary Celeste.

It will only be those cities which have been abandoned not in a controlled and orderly fashion, but in turmoil, whether by flood, or fire, or earthquake, or with its final days contested by military factions and the death squads, that human remains may be reasonably common. In much this way, the volcanic ash raining down on Pompeii first trapped and then buried some thousands of Roman people, for archaeologists to excavate nearly two millennia later.

The volcanic ash here preserved not just the skeletons. It moulded around the bodies to preserve the external surface—a geological rarity that allowed features of the skin and external musculature of the unfortunate Romans to be made once more visible, by the simple expedient, devised by Giusseppe Fiorelli in the nineteenth century, of injecting plaster of Paris into the moulds. Such preservation is marvellous, if a little terrifying as a *memento mori*. But, this type of fossil might not survive over truly geological timescales, even if Pompeii itself, or a Pompeii-like city of the present or future, were to be located on a downgoing escalator, and become buried by more strata. For the pumice particles that make up the ash cover at Pompeii are easily compacted and flattened, once the overlying strata begins to pile up above them. The delicate gas bubbles in the pumice, and the larger corpse-cavities where humans had lain, would both be flattened and distorted, perhaps beyond recognition or reconstruction.

When human skeletons are unearthed, it will likely be in numbers, and in geometrically arranged rows, each skeleton separate, the whole being geographically close to, but separate from, the constructions of the Urban Stratum. The human cultural practice of burial is widespread. Not universal, of course. Cremated bodies are self-excluded from fossilization—though, maybe, not completely. The small carbonized fragments that remain may preserve the microscopic structure of bone better than can be found in more complete and unburned skeletons. Palaeontologists today who study fossil plants often look in particular for fragments of charcoal produced in ancient forest fires; here, the charcoal represents plant material that has been reduced to a carbon husk inedible even to most bacteria, and this husk can show marvellously clear cellular structure, to a microscopic level, in three dimensions.

For a future palaeontologist to draw similar conclusions from fragments of cremated human bone, though, would involve some painstaking taphonomy and inspired scientific deduction. First, finding the tiny bone fragments and recognizing them as such, then telling them apart from non-human animal bones burnt for human predation and consumption, then deducing burning as a widespread human practice, post-mortem. This facet of human culture—and hence this aspect of human skeletal micromorphology—thus seems unlikely to be reconstructed.

On the whole, even without charring, animal bone preserves reasonably well over the geological long term. It is, though, a multifaceted substance: as well as mineral, it contains proteins, blood vessels, nerves, and 'hard' and 'soft' cellular tissue components. In life it is dynamic, having to grow as the animal that it supports grows. And it is complex: within it, concentric mineral layers enclose bone canals (or more correctly Haversian canals, named after one Clopton Havers) that contain blood vessels. Normally the interior, where resides the marrow, is spongy, especially in animals (such as birds) that need to combine strength with lightness, while more compact bone forms the external parts. The bone cells themselves occupy tiny cavities between the concentric mineral layers. Some bone is toughened and highly resistant, notably the enamel-reinforced grinding surfaces of the teeth, that need to survive a lifetime's grinding and chewing.

Bone is by no means indestructible, though. On the surface, or shallowly buried in sediment, it is prone to attack by insects such as carrion beetles (given the lovely term 'ostrophagous', meaning bone-feeders), that excavate distinctively shaped chambers in the bone for their larvae, or carve meandering feeding traces into its surface. Microbes, too, attack the bone (even living bone, to the extent that dentistry is a profitable profession). Electron microscope pictures of fossil bone commonly show networks of tunnels, each just a few thousandths of a millimetre across, ramifying through the bone material. Left thus to the elements and the ostrophages, bone can begin to disintegrate within weeks.

Long-term preservation of bone typically involves mineralization. In the burial realm, bone cavities can fill with minerals precipitated from coursing fluids—calcite, or silica, or pyrite, or iron oxides, depending on the chemistry of those fluids and how they interact with the bone material. The course of fossilization can vary greatly. Take the Taung skull, for instance, all that remains of the poor child (the prey of an eagle, it is said) that now has the responsibility of representing the species *Australopithecus africanus*. This was blasted out of a South African cave, embedded in calcite-cemented cave deposits as hard as concrete. It took much painstaking, careful excavation to free the bones from their protective rock matrix. The bones of *Homo floresiensis*, though, when found in their cave in Indonesia, had the consistency of mashed potato. It took considerable luck, and then immensely gentle handling, for those fragments to survive their discovery and excavation. But then, hobbits are lucky by nature.

THE SOFT FACTOR

A skeleton is an eloquent assortment of bones, but it is not a body. How could our excavators reconstruct a more *rounded* outline from those bones, give humanity a human face? Future palaeontologists would need to do this on the basis of anatomy studied, learnt, intellectually absorbed, from whatever living mammals there will be on this Earth one hundred million years from now: placing internal organs, nerves, and blood vessels where they should sensibly go, adding layers of muscle, meshing them with the reconstructed bones into a functional system of supports and levers, then adding layers of fat and skin and connective tissue and fur or scales. This is basically how vertebrate palaeontologists proceed today to reconstruct dinosaurs and the strange mammals of the Tertiary Period. But this approach will not predict everything. Getting the nose and ears right will be tricky (no external ears at all would be one possibility, or rabbit-sized ones another), while other options might be mooted, such as whether or not we had a sail on our backs or a crest on our heads.

So could any human soft tissue be preserved, either as impressions or as some petrifaction of the flesh itself, to anchor the reconstructions of the future scientists? Soft tissue preservation in the fossil record is a rarity, and localities where it occurs are given their own name—*Lagerstätten*—by palaeontologists, who hunt for such sites tirelessly. Once found, these *Lagerstätten* become celebrated as rare and extraordinary windows into the biological riches of the deep past.

Some of these windows only stay open for a short time, though, before their inevitable and eternal closure. The frozen mammoths of Siberia, preserving skin, hair, gut, eyes, will decay once the permafrost melts, as would almost certainly happen over timescales of tens of millions of years anyway (even if humans had never evolved nor the Model-T Ford been invented) as continents shift and Ice Ages give way to global greenhouses. However, if the present warming of the world is at the mid to high end of the range of predictions, then the ground-ice will vanish more quickly: most of the permafrost, together with its treasure-chest of mammoths, will be gone just centuries or millennia from now. The mummified giant sloths of the arid mid-west of the USA, likewise, will become normally decay-prone when the climate in that part of the world eventually turns humid. Both mammoths and sloths, too, are subject to the paths taken by whichever tectonic escalator (being on land, this would be mostly upwards) they happen to be on.

More durable *Lagerstätten*, giving snapshots of life in the round going back half a billion years and more, include rock strata such as the Burgess Shale of British Columbia (preserved itself in the *Wonderful Life* of Stephen Jay Gould), the Soom Shale of South Africa, lying high up on Table Mountain, the Rhynie Chert of Aberdeenshire, the Solnhofen Limestone of Germany, the nodule bed of Herefordshire, the Chengjiang deposits of China, the Santana Formation of South America; and others still. Each reveals details of the soft parts of animals, sometimes down to microscopic, almost molecular levels. The fossil fish of the Santana not only preserve their muscles, but they preserve, flash-petrified in calcium phosphate, the bacteria that began—but didn't finish—the action of decaying them. The Rhynie Chert contains some of the earliest aphids, some 400 million years old, while the early plants that they fed on still preserve the puncture marks made by those hungry prehistoric greenfly. And the Herefordshire water-fleas, of similar antiquity, are so well preserved that they reveal that the males of the species possessed, as do their modern descendants, a penis as long as the animal itself, a discovery that excited considerable tabloid interest at the time.

Most of these extraordinary deposits were formed under water, and most were under seawater. It is their rarity that links them, and the speed of the fossilization (with petrifaction taking place in hours rather than over millions of years). But there is no comparable uniformity of preservational mechanism. It is this variety of process that gives some prospect that some vestiges of proper human shape might, just possibly, accompany the bones that we will leave behind us.

The Rhynie Chert was almost a dry-land deposit, being a peat bog that formed next to the hot springs of a volcanic system. The plants, and the insects that fed on them, were suffused with the silica that precipitated from the hot water around them and within them and became their tomb. The puzzling thing here is that hot springs are common, and silica precipitation generally is downright commonplace (producing varieties of chert such as the flints that are abundant within the Chalk)—yet the Rhynie Chert remains as a justly renowned singularity. The devil here must lie in the detail of the nature and timing of how the silica was released into the system and just how it invaded the tissues and precipitated within them, as crystals consistently fine enough to preserve details of cell structure. Grow the crystals too late, and there will be nothing to preserve. Grow them too big, and the crystallization process will wreck the cells, not preserve them.

Out to sea—just—was the Solnhofen Limestone, famous for preserving the Jurassic *Archaeopteryx*, feather impressions included, in a lithographic limestone. Henri Toulouse-Lautrec admired this particular rock type. It is so smoothly and finely and evenly laminated that he could draw freely on it with a grease-based crayon (or 'tusche'). The limestone surface was then wetted, and an oil-based ink applied, that clung to the crayoned surface but was repelled from the wet areas. Pressed to paper, the drawing could be faithfully transferred, and the process repeated many times, for the limestone (unlike copper engraving plates) did not wear out. As a child, I was both fascinated by fossils and had a copy of Lautrec's *Aristide Bruant* poster on my wall, with no inkling of the strange link between the two enthusiasms.

The Solnhofen Limestone was originally a lime mud, accumulated on the floor of a submerged hollow, separated from the open sea by coral reefs. Most lagoon muds are home to countless burrowing and chewing organisms, that would make short work of any animal that fell in, even if the animal were as distinguished as the *Archaeopteryx*. Not this lagoon. The layers piled up evenly and consistently—so consistently that quarrymen can still recognize and trace individual layers, less than a centimetre thick, between quarries that may be several kilometres apart. The lagoon was obviously hostile, and it is speculated that it was sufficiently cut off from the sea to have become hypersaline, the densest and most poisonous brines lying on the lagoon floor. The striking 'death march' fossil of the horseshoe crab, preserved at the end of its brief trail, suggests this, as do some of the fossilized fish that are contorted, seemingly petrified in the act of thrashing about. The fossils are impressions, caught in the rapidly hardening fine mud as nicely as, much later, Toulouse-Lautrec's fine etchings were fashioned in preparation for printing.

The Solnhofen rock is exceptional—but it is not a bonanza, at least as regards quantity. Magnificent fossils are thin on the ground. The fossils were only found because of the large quarries dug into them (and particularly when the quarrymen who extracted the limestone worked by hand, slab by slab—and were paid extra for any particularly striking fossil). Search now among the slabs and you will be lucky to find anything that resembles the much-photographed museum specimens—with one exception. The limestone surfaces abound in *Saccocoma*, a beautifully preserved stemless crinoid or 'sea lily', arms entwined elegantly around its crown. *Saccocoma* may have been a bottom dweller, tolerant of high salinities, but is more widely considered to have swum in the surface waters, thus keeping sensibly clear of the lethal brines below.

The Herefordshire deposit was further out to sea. It was a Silurian sea floor upon which fell a layer of volcanic ash. This entombed worms, sea-spiders, those well-endowed water-fleas, and yet other creatures and then, very shortly afterwards, calcium carbonate precipitated around the animals in the ash as walnut-sized concretions, before the creatures had a chance to decay or to be flattened by compaction. Now, ash layers within strata generally are common, ten a penny, and may appear in their dozens in the rock face of a cliff or quarry, interlayered with the normal muds or sands or lime deposits that formed on that sea floor. So why are those not full of beautifully preserved fossils? As with the Rhynie Chert, it must be because of some quirk of mineralogy or chemistry (or both) of that Herefordshire ash. Some interaction of ash and seawater and sea floor sediment concatenated to release those calcifying solutions and crystallize nodules from them, almost as fast as lightning. This doesn't happen every day.

Then perhaps the most famous of all, the Burgess Shale, though it is in some danger of being eclipsed by its near-contemporary, the Chengjiang mudstones. Both are Cambrian (the Burgess deposited mid-way through that period, the Chengjiang early within it); both represent muddy sea floors, though the organic-rich Burgess rock is almost black, while the organic-poor Chengjiang is a pale straw-yellow. Both contain some of the earliest animals to appear on the Earth: sundry worms and arthropods, mostly, plus a sprinkling of exotica, including the first vertebrates (at both sites) that include genuine—and unexpected—fish (at Chengjiang) and also some bizarre creatures—early experiments in multicellular life, perhaps—that remain mysterious still. These menageries come complete with guts, skin, nerve canals, blood vessels, eyes as thin carbon films, often embroidered with a tracery of microscopic pyrite crystals (or, in the case of the Chengjiang, of iron oxides that have remained after surface weathering of the original pyrite).

Here again lies an enigma, for marine mudrock is the commonest sedimentary rock on this planet, and many mudrocks contain nice fossils, to be sure, but it is only a very, very few that contain such extraordinary fossils as two-dimensional carbon-and-pyrite ghosts. Was the sea floor stagnant, anoxic, like the floor of today's Black Sea?—for lack of oxygen will prevent scavengers from living on a sea floor and eating any carcasses that fall into it, and also slow the bacterial decay of animal tissues. The Burgess Shale certainly approximates to a black shale (though the Chengjiang does not), but the jury is out on whether the sea floor it represents was truly stagnant, for sporadic burrows have been reported within it. And, in any event, anoxic sea floors and the formation of nicely laminated, unburrowed black shales commonly took place in the past, notably during the Palaeozoic Era.

Such black shales abound in some of the older mountain ranges—the Welsh and Scottish hills, the Appalachians, the Ardennes massif, the Sudetan mountains. They preserve some wonderful plankton, to be sure: but almost always plankton that had some form of resistant skeleton. One can find elegant graptolites by the bushel, sometimes the remains of thicker-shelled planktonic crustaceans, sometimes the bony 'teeth' of the conodont animal (once enigmatic, this particular creature is now unmasked as an early, sardine-sized and eel-shaped vertebrate with entirely non-standard dental arrangements); and microscopically, in their millions, the spore-like resting stages of marine single-celled algae. But little else—and, heaven knows, enough palaeontologists have pored over such rocks in search of fossil bonanzas. Of most of the plankton, that would have teemed then, we have no idea. The past empires of the medusae and of their relatives the comb jellies, of the thin-shelled crustaceans, of the salps and arrow worms and planktonic polychaete worms are essentially vanished realms. We have little hope of ever reconstituting them. They dominated the Palaeozoic oceans, but their history looks set to remain forever cryptic.

Why so? For such stagnant sea floors should, one might have thought, act as preserving traps for anything that fell in, whether hard-shelled or soft-skinned. But, almost always, they did not. One factor here is the recycling of dead matter back to the living, in the oxygenated upper layers of the sea. Open ocean waters often approach being deserts, as far as nutrients are concerned. Any dead organism is thus the equivalent of manna in this desert, and something will try to recycle it, particularly as the decaying tissue may initially be gas-rich, and so take some time to sink. Nearer shore, where nutrients are more plentiful, the chances of actually reaching the sea floor are better; there is still intense recycling, but the long, slow descent may be speeded if organisms are caught up in sinking marine snow, a material made up of gelatinous debris such as discarded feeding nets of tunicates and the faecal pellets of planktonic organisms. The faecal express, it has been called, as a means of denoting the rapid transit of matter from sea surface to sea floor. On the sea floor itself, even if anoxic, there are still microbes in abundance, often forming complex mat-like structures. These are the analogues—and many will be the descendants—of the microbes that ruled the early Earth, bereft of free oxygen. They too will use infalling animal tissue as a source of energy and materials, albeit with less gusto and mechanical vigour than organisms of the oxygenated realm.

Indeed, given this rich microbial life, even on anoxic sea floors, one can marvel at the preservation of such fossils as the graptolites. Their skeletons are organic, being made of collagen, similar to the stuff your fingernail is made of. Where sediment accumulated only slowly—a few millimetres a century is typical—such a skeleton must have lain at or just below a sea floor for centuries or millennia, shrouded in microbial mats (a vision worthy of one of the more atmospheric horror films, like some miniaturized prehistoric *Nosferatu*). Yet the fossils, excavated many millions of years later, show wonderful detailed structure. It is a tribute to the resistance, or perhaps the inedibility, of their skeletons.

What was the secret, then, that governed the sleight-of-hand that removed the Burgess and Chengjiang organisms from the recycling mill? These strata are both open-sea deposits (the faunas in each case represent fully marine conditions), but were relatively close to shore. The palaeoenvironment of the Chengjiang has not been fully resolved (it is a relatively new find, and attention is still focused on the amazing animals it contains). The Burgess Shale, as an object of scientific enquiry, is longer in the tooth and so has been more widely studied. It accumulated as muddy layers at the bottom of a steep, almost precipitous underwater cliff of limestone. The layers themselves do not much resemble slowly accumulated graptolite shale, that piled up with infinitesimal slowness as flake after flake of mud drifted down through still ocean water. Rather, the Burgess strata seem to have formed as slurries, that sped down the cliff slope and piled up as layers, perhaps up to some metres thick, around its base. The animals were either carried down with the mud or were buried in place (quite where they originally lived within that submarine geography seems unclear).

In any case, the Burgess carcasses were taken, almost instantly, below the realm of burrowing and scavenging and of intense bacterial action. They were quickly flattened, the soft sloppy mud that entombed them becoming compressed into shale as more sediment piled on top. Over time they descended to perhaps some ten kilometres below the surface, and were heated to some 300 degrees centigrade. The carcasses lost hydrogen, oxygen, nitrogen, became essentially wisps of almost-pure carbon, but remained intact. Then came the building of the Laramide Mountains in late Cretaceous times, and their escalator changed direction. The immense forces squeezed most of the surrounding mudrocks into slate, reorienting and recrystallizing the mudrock minerals. Had that happened to the Burgess Shale, the rock would no longer have split along the original lamination, and the fossils would have been effectively impossible to find. Once again, though, the Burgess animals had a charmed afterlife. The massive limestone cliff acted as a buttress, and the Burgess Shale, in its pressure-shadow, was spared from transformation into roofing slate and billiard-table. Its escalator continued upwards, helped by the insistent pressure of the Pacific Ocean crust against western North America. In the geologicaly recent past, it came to the surface on Mount Stephen, in British Columbia, the shape of which had itself been sculpted by the great Laurentide ice sheets. A century ago, Charles Walcott discovered it, and wrote a notable page in the history of life (a page subsequently embellished beautifully by Harry Whittington and his students, and advertised eloquently by Stephen Jay Gould). And, geologically soon, this particular window will close, as the weather eats the mountain, and the Burgess fossils will be gone forever. *Sic transit gloria mundi.*

EXCEPTIONAL HUMANS

So, given that there are a range of means of preserving animals exceptionally, and not just ordinarily, could a human soft-tissue *Lagerstätte* arise, and smuggle the likeness of a person into near-eternity? This is unlikely to happen far out to sea, preserving pirates, say, who were made to walk the plank: Davy Jones's locker holds too many means of recycling human chemistry for that. Nor on land, for like reason, and because of the inexorable action of the ascending escalator. The preserving bottle, if there is to be one, will likely be somewhere around the thin line that separates these two domains, somewhere around the coastal lakes and lagoons and deltas and estuaries that lie between land and sea, and where the sediment piling up is pressing down on the pliable crust. There must be rapid burial, to confound the scavengers, and then chemistry should act on the flesh exceedingly quickly, before the microbes (no slouches, they) can carry out their work.

In its collective actions, humanity may have stacked the deck in its favour, as far as this particular detail of eternity is concerned. Burial is helped by our cultural practices with the dead, though in most cases one might suspect that the separation of body from sediment (and hence from its potential petrifying action) by coffin or sarcophagus would hinder the long-term petrifaction of flesh, even while helping the preservation of bone. It is during accidental burial that favourable circumstances might arise—though the term 'accidental' here should perhaps be broadened to include the effects of warfare and other forms of murder.

Here, one of the geological effects of humans, as eroders of landscape and producers of sediment, may come into play. The enhanced and accelerating deforestation of most terrestrial landscapes, widespread ploughing, the reshaping of landscapes for building, and the extraction from that landscape of the materials for that building—all these have increased, perhaps as much as tenfold, the amount of sediment being moved at the Earth's surface, compared with an average baseline state for that surface. And with so much extra sediment around, much of it unanchored by vegetation, there is that much more scope for landslides and mudflows, that can periodically (as one can see in the headlines, from time to time) engulf settlements and people, burying both in a new catastrophic event layer.

It is not quite so simple, of course. Much of the sediment that is carried by rivers is now stored in the reservoirs behind the dams that hold up the flow of most of the world's large rivers. This makes these reservoirs quite temporary as structures of use to people (they will silt up in a few centuries), and they also keep the sediment from reaching the coastal deltas. The Nile delta, thus starved, is currently denied the sediment that would fertilize the crops and keep building the delta (that is subsiding under its own weight) above sea level (which is encroaching ever more on to the land). The sediment layers trapped in the reservoirs (and any fossils, human or otherwise, that they contain) will be recycled back to the sea, once the structure is abandoned and the dam is eventually breached. This will destroy all but the most resistant of the lake-held fossils in the process.

Nevertheless, *in general*, a destabilized land surface, allied to a human population explosion (moreover, one that is concentrated in the coastal zone of maximum fossilization potential), should increase the chances of the kind of events where the processes of exceptional preservation might come into play. The newly warmed world, too, may see more of the kind of tropical storms and hurricanes that act as triggers to landslides and mudflows. This is disputed, though: there will be more energy in the weather systems, yes, but that might not translate into more powerful hurricanes, because the extra energy could tend to tear the hurricanes apart rather than help them grow to full strength.

There may be novelty in the chemistry of fossilization, too. Human society has been carrying out some large-scale experiments with environmental chemistry that are brewing up some interesting, palaeontologically active concoctions. For instance, there has been nourishment of the waters of many rivers, estuaries, lagoons, and coastal bays, a by-product of the immense efforts made, year after year, to nourish the world's growing population. Phosphates and nitrates are pouring off agricultural land into these waters at geologically unprecedented rates; the world's nitrogen cycle, for instance, has been roughly doubled by human activity. These chemicals stimulate the growth of marine planktonic algae that, in decaying, use up the available oxygen in the water. Eutrophication, it is called, and it has already created enormous annual 'dead zones' that cover some hundreds of square kilometres, such as in Chesapeake Bay in the eastern USA, where bottom-dwelling creatures are killed off wholesale. This shuts off scavenging action, while the alternation of oxidizing and reducing conditions helps bring the sulphur-reducing bacteria of stagnant muds into contact with the dissolved sulphate ions present in seawater. The ensuing crystallization of iron sulphide as pyrite around organic remains is a prime candidate for initiating exceptional preservation of fossils of the Human Age. The eutrophically enhanced phosphate levels, likewise, cannot harm the chances of phosphatization of animal tissues in such zones.

Other types of chemistry may come into play. The wholesale transfer of limestone from inland areas into the concrete edifices of coastal zone cities has brought massive amounts of calcium carbonate into the prime zone of long-term fossilization. Vary the pH of the water via an assortment of urban effluents—sometimes acid, sometimes alkaline—and put this into contact with animal remains. The scene is set for reactions involving the dissolution of calcium carbonate from the concrete (and from limestone used as building stone) and then its precipitation as concretions around the animals. There are a whole variety of other chemical effluents that come under the general heading of industrial pollution: some metal-rich, others charged with hydrocarbons, some as brines, others with sulphates or chlorides, often concentrated in dangerous, unfenced industrial lagoons and sludge-pits set in the middle of teeming humanity. We collectively may be unconsciously inventing a whole new set of scenarios for exceptional fossilization.

Then there are cultural factors, that may inadvertently lead to some quite effective soft tissue preservation. One example is the fabled practice among the gentlemen of the New York Mafia of permanently terminating the ambitions of business competitors and then dropping their mortal remains into Hudson Bay in concrete overcoats. This practice creates moulds of the body outline (with mantling trench coat, fedora hat, and spats, perhaps) in a geologically hardwearing material, while placing the remains directly into an environment where scavenging is limited (the weight of the concrete should sink the cadavers some distance into the mud) and where pyritization and phosphatization of soft tissues can instantly begin. It is hard to say how widespread this business practice is (one of the drawbacks of an academic life), but if repeated with any frequency along the seaboards of the world, it might be producing some interesting taphonomy. Thus, crime might pay, eventually. At least as regards future science.

There is the human impulse, too, to make images of ourselves: pictures, paintings, drawings, films, statues. Can any of these survive, to be linked with the skeletal remains and tissue fragments, to add to our future reconstruction?

Most of the modern means of illustrating people seem to have rather poor preservation potential. Anything on paper or canvas would almost certainly be carbonized beyond recognition over the time span we require. Anything electronic, as hard or compact discs or magnetic tape, or reels of film, or the delicate equipment designed to play these, would likely corrode or dissolve, as would metal engravings, beyond, at least, effective analysis and interpretation. This seems to leave perhaps the oldest means of all: the making of statues in stone. These are, at least, in geologically proven materials as far as longevity is concerned. Let us hope, though, for the sake of a proper grasp of human morphology, that the first statues to be unearthed are of the natural or socialist-realist school and not abstract, conceptual, or minimalist.

FROM THE HEAVENS?

It seems reasonable to predict that human skeletons—and perhaps fragments of preserved soft tissue—will turn up, in enough numbers and in enough proximity to the remarkable trace fossils of the Urban Strata, to make it clear that humans were the species behind the material civilization that appeared so dramatically upon, and then disappeared so quickly from, the Earth. But that discovery would open the enquiry, rather than close the case. For, piecing together the links between the evidence of geologically abrupt climate and environmental change, the vast engineered cities, and the fossilized remains of humans may leave one outstanding question: where did humans come from?

Did they indeed originate on Earth, or did they represent some ill-starred interplanetary settlers, a colony that grew out of hand? The human skeletons will be sufficiently unlike the other animal fossils of the time to make their origin and their affinity uncertain, and an extra-planetary origin might well be tossed into the hatful of competing hypotheses for this palaeontological phenomenon.

Such an idea would not be ridiculous, for the human takeover of Earth was, in effect, accomplished in perhaps 10 000 years, and that interval, at a distance of 100 million years, will appear virtually instantaneous. Furthermore, the takeover accelerated almost exponentially near the end of that brief span, in that humanity has produced more body and (especially) trace fossils and environmental ripple effects in the last few centuries than in all of the preceding part of that span. At such a remove, this will appear quite as sudden as the meteorite strike that killed off the dinosaurs. Amid the ruins of the Urban Strata that hugely expanded in the last century, it may be hard to find signs of cultural evolution in the preserved rubble; in any event, the distinction, in alien eyes, between, say, Roman bricks and tiles and Victorian or modern ones may be an archaeological subtlety too far. The stratigraphic evidence will likely suggest instant arrival of a civilization that was fully formed.

The intergalactic explorers of the far future will also find it much, much harder to find fossils of our primate ancestors than do human palaeoanthropologists. And, goodness knows, such fossils are hard enough to find today, even with a wealth of suitable terrestrial strata—the Olduvai deposits, for example—near the surface, and the caves formerly inhabited by Neanderthals and hobbit-people still present, still open and still ripe for exploration. A hundred million years on, most of this evidence of human ancestry will be gone—and most of the small fraction that will remain will be deep underground, covered by later strata. To the future explorers, humankind may really appear to have descended, in its landscape-refashioning multitudes, from the skies in a kind of blitzkrieg.

Any emergent ideas of humans-as-space-travellers should, though, be dispelled by studies of comparative anatomy, as our chroniclers get to grips with the classification and relationships of the myriad organisms, past and present, of Earth. The relations of the human species to other vertebrate species can be established by study of their skeletons, in much the same way that Gideon Mantell established the similarity between the dinosaur teeth that he found in the Sussex Weald and those of the iguana, calling his fossil animal the *Iguanodon* (being persuaded by his colleague William Conybeare out of calling it the *Iguanosaurus*). Humans share with other vertebrates a bilaterally symmetrical internal skeleton made of calcium phosphate, including a backbone, four limbs, and a cranium, albeit a bizarrely expanded one. Assuming that our future explorers have not (by some kind of interplanetary convergence of skeletal morphology, biochemistry, and mechanics) evolved a similar skeleton, that should indicate the generally Earthly affinities of the human race; such a demonstration of relationship, though, will likely not (assuming human-like levels of curiosity in our excavators) put a stop to discussions of human origins.

For the differences between humans and other fossil mammals that are likely to be preserved will be striking. It is likely that, outside of the narrow event-stratum representing human occupation, there will be nothing quite like us—at least nothing that will be easily found by those geologists of the future. The fossil record of mammals that will be available at that future point will be more like the preserved record that we have of the dinosaurs, than it will be of, say, the range of Ice Age mammals that we can now excavate. Of today's beasts, those that will be preserved will be common herbivores that live on low-lying deltas and swamps above a downgoing tectonic escalator. Mammals inhabiting higher ground (including most non-human primates, which tend to be relative rarities anyway) will have little chance of getting into the fossil record. So humans will not be neatly fitted into some family tree constructed from recovered fossils of different times in the way that has been possible for fossil horses. Thus successive primitive horses—*Hyracotherium* (more popularly known as *Eohippus*, or the dawn horse, but the 'hyrax-like beast' has technical priority: the rules of taxonomic nomenclature are strict, and take no account of poetic qualities); *Mesohippus, Merychippus, Pliohippus,* and their kin, with their trend of increasing size and reduced numbers of toes, can plausibly be fitted into lineages that have led to *Equus*, the modern horse with its single toe-as-hoof, (and also to the zebra, the ass, the quagga, and others on different branches of that tree). Humans, by contrast, will seem a puzzling singularity.

What will there be to compare the fossilized human bones with? There will be the bones of other animals in the almost infinitesimally brief interval of human domination—a tiny fraction of a million years—in which the mammal population has been hugely biased to exclude and eliminate competitors. Here, there will be the remains of those animals that are cultivated in order to be predated more effectively: cows, sheep, pigs, goats. Some animals—dogs, cats, horses—have functions today that will likely be indistinguishable from predation on the basis of available evidence. Then there will be those animals that lived in geologically immediately pre-human times: say those going back some few million years to encompass the time interval that we now assign to the Quaternary (i.e. Ice Age) and late

Tertiary periods. Quite a few of those animals are larger than humans—assorted ungulates, deer, elephants, hippos. There may be remains of the more successful carnivores: fragments of lion, tiger, hyaena, wolf: all creatures that used to be more common and widespread before they were elbowed out of the way. There is little room for such competitors on a planet where a quarter or more of the planet's entire productivity is feedstock to grow more humans with. There will be some smaller creatures, too, as fossils: rodents, in particular, one would guess. These are known as fossils mainly from their resistant teeth. These are almost all basically quadripedal, keeping all four feet more or less firmly on the ground.

And then came this brief population explosion, of an animal with an exaggerated, thin-walled cranium; with delicate hands and fingers (only to be found fossilized in the best specimens); with small, generalist teeth—hardly those of a typical top predator—and no claws, nor hooves.

Would it be reconstructed as that rarity, a bipedal mammal? Perhaps, once the explorers developed a good enough grasp of the skeletal mechanics of hips and of vertebrae and skull–neck connections. As largely naked? Probably not, at least not from the evidence of the skeletons. Any palaeontologist worth his/her/its salt would note (once enough fossils turned up, at enough localities) the extraordinarily wide latitudinal range of distribution, and the association with cold as well as warm palaeoclimates, and reconstruct this strangely adaptable being to include a good coat of fur (until those tell-tale statues are unearthed, and fur is replaced by a puzzling variety of temporary body surface coverings).

Even with anatomical links established, firmly linking the fossilized humans with other terrestrials, the enigma of human origins would surely persist. A population explosion obviously—but of a creature with no obvious direct ancestors, and one that seemed to have evolved its civilization virtually instantaneously. In the position of a visiting inter-galactic explorer of highly advanced technology, one might suspect that previous inter-galactic explorers had visited long before, and had carried out a little speculative genetic engineering to speed up the development of an indigenous intelligent civilization on a living but untutored planet. And that they had bungled the job.

Far-fetched? Another alternative might seem, at such a distance and with such evidence, just as unlikely. That there had been a succession of relatively rare highland mammals, which entered the long-term fossil record so rarely as to become effectively invisible. That they were originally descended, perhaps, from some kind of insectivore, and that this undetected lineage had long been evolving on the fringes of a highly developed terrestrial ecosystem. Then, that evolution in one small population had accelerated, or reached some kind of threshold, where the unusual combination of features, of expanded cranium and upright stance and manipulative ability, had given some kind of instant competitive advantage over every other creature in the planet, even over much bigger and fiercer ones. That the population had then increased at an almost bacterial rate. And that almost immediately afterwards, geologically, it had disappeared simultaneously with major climate upheavals and a more widespread wave of migrations and extinctions throughout land and sea.

The manner of the disappearance must spark equal controversy among our future scientists. Did this remarkable species have the utter bad luck to make its evolutionary appearance just as a major climatic and environmental reorganization of the planet was beginning, and perish in the upheavals? Or did it combine high intelligence with breathtaking stupidity in equal measure: to be able to dominate the environment on the one hand and create a technologically sophisticated empire, but simultaneously to dismantle the systems that kept the Earth's surface stable and habitable?

Such a question is an almost exact parallel of the current debate over what killed off about half of the world's large mammal species—including mammoth and auroch, giant sloth and woolly rhinoceros—just a little over ten thousand years ago. Rapidly changing environment at the end of the last Ice Age (including that brief freakish reversal into the bitterly cold thousand-year spell of the Younger Dryas event, a millennium after warming had started)? Or overkill by an expanding and migrating human population? The jury is still out (the latest models suggest a complex scenario in which humans and environment both played a part)—and this is after intensive research into an extinction event that took place geologically almost yesterday, with abundant near-surface evidence and a chief protagonist (the human species) whose capabilities and motives we are very familiar with.

Make such an event much wider in scale and complexity, in the form of the slightly later Sixth Extinction (as currently unrolling). Translate it back a hundred million years, with all the blurring and fragmentation of evidence that that entails, and then place it on an unfamiliar and newly discovered planet. Suspect a central position in this Earth-scale drama for a remarkable, distinctive bipedal species that apparently lacked close relatives but constructed vast aggregations of artificial stone. The nature of this species would come under the microscope, as a phenomenon that did more to the Earth than build city-structures on it. From the evidence, then, would our future explorers be able to tell whether this remarkable ape was brute or angel?

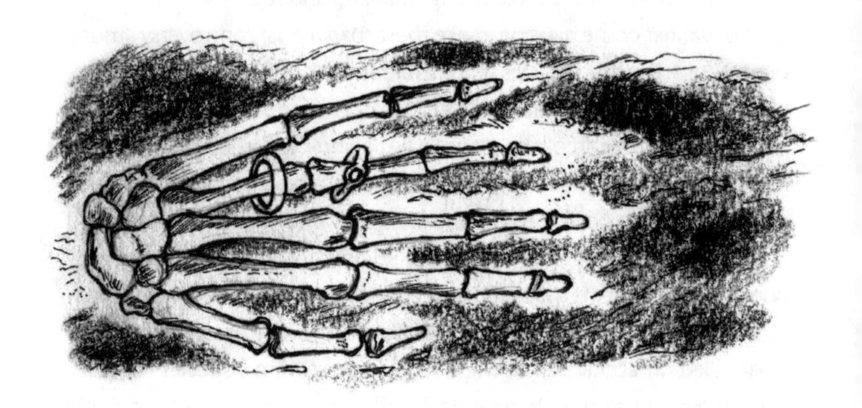

Meeting the People

It would be wrong to say we understand these long-dead city-builders, who in some senses are our predecessors. Some of their relics seem almost familiar, but there are indications also that we find unsettling. The evidence of their behaviour is ambiguous, and much is infuriatingly contradictory. What drove them to create what they did—and what killed them? We cannot help speculating on these extraordinary beings, even as we realize that much of this speculation is irresolvable and hence fruitless. Their extinction is a great pity. For all the possible dangers of contact with a species associated with such planetary instability, with such changes in their habitat, what might we have learned about them—and from them—had they survived?

If one has found the ruins of an ancient civilization on a distant planet, and the skeletal remains of some of the civilization-builders amid those ruins, there would follow a flood of questions—if the discoverers possess anything like human levels of curiosity, that is—about just what these beings were like. War-like and aggressive, or peaceful and harmonious? Socially cohesive and communicative, or individualistic? Sedentary or highly mobile? Rational or superstitious? Good or bad . . .

What could an impartial observer, coming fresh to such a scene, make of humanity's actions and habits and motives? It is hard—well, yes, impossible—for us to think through the mind of an extraterrestrial alien. It is hard enough to think through the mind of another human being. In the study of past humanity one might take the example of Stonehenge, a magnificent, geometrically highly ordered structure, designed and constructed by members of our own species with great ingenuity and effort. It served, quite obviously, a highly important purpose. But just what was that purpose? All we have is the structure itself, and the archaeological evidence that surrounds it (for there is a stratigraphy too in such studies, of placing physical evidence within a time framework). We have lost all contact with the constructors, for the chain of word-of-mouth history-telling has long been broken, and if any kind of written records were ever produced, none have been unearthed. But this is an artefact of our own species, remember, and there are many potential parallels, in modern and ancient culture, that we can draw on.

Yet basic questions about its function continue to be debated. Was it produced for ritualistic purposes, from deeply felt religious motives? Or for purely practical purposes, to do with the organization and management of goods and labour. Or as an awe-inspiring symbol of naked political power? Did the people of those times go there as people go to church today, or as they go down to the pub, or to a wedding, or to the labour exchange or to Buckingham Palace or perhaps to a village council meeting? Or perhaps they went for different reasons at different times: the structure was built and embellished over something like a thousand years, so its function and context may well have evolved. The construction was designed with insight into astronomy, the stones being aligned with rising and setting points of the sun and moon; but, again, was this largely practical (to do with the timing of the seasons for agricultural purposes) or religious (as an expression of awe at and subjugation to the surrounding forces of the Universe)?

The debates will probably go on for as long as there are archaeologists to argue over the evidence, whether in lecture theatre or meeting convivially afterwards, pint of beer in hand. The debates over the fossilized, physical remains of an alien humanity (alien to the intergalactic explorers, that is) and its constructions will undoubtedly range over a greater diversity of possibilities, real or imagined. Still, one can try to approach the exercise as might some kind of combination of Sherlock Holmes and Mr. Spock, attempting to draw rigorous, logical inferences from the evidence, rather than as an Inspector Maigret, divining the psychological motives of the main characters. (To analyse some aspects of humanity, the perspective of a Groucho Marx may, of course, be more appropriate.)

Some basics, first. How many species of humans? A re-run might be expected of the dilemmas of palaeoanthropologists, extracting fragmentary skeletons of early hominids of different ages and from different localities and then debating over whether they represent several closely similar yet distinct species (the 'splitting' tendency in taxonomy) or one rather variable species (the 'lumping' tendency). Palaeontologists working on any fossil group, in fact, are all to some extent lumpers or splitters, as all they have to go on, normally, is the remains of hard parts: no soft parts (normally); no genetic composition; little of ecology and even less of behaviour. And the biologists among the future explorers will likely by then have been grappling with the concept of Earthly animal species and of intra-species variation.

Future human fossils will appear to be confined within the strata of one geological time instant, so that will at least be one constraint on the possibilities. However, the geographical spread will be greater than that for any other organism, living or fossil; with enough searching, remains should turn up on every continent, with the likely exception of Antarctica, that, a hundred million years on, will likely have drifted into higher latitudes, and be green, and not ice-bound. (An astute future stratigrapher might note here the connection between an enormous gap in the stratigraphic record there associated with evidence of glacial erosion at the time level of the fossilized cities, recall the necessity to have large amounts of ice somewhere on the Earth at that time to generate the large rapid sea level changes of that time, use palaeomagnetic data to place Antarctica directly over the South Pole, and correctly reconstruct that continent as ice-bound and so generally uninhabitable, even for humans.)

Some aspects of current human variability will not be preserved—eye and skin colour—while others will: body length, as say between Kalahari bushmen and the Masai of Africa. It seems reasonable, in this case, that the general similarity of all human skeletons to each other (variation within humans does not exceed that generally seen in other species), and the striking difference from contemporary vertebrates that are likely to be preserved, would lead to the simplest conclusion: just one species.

Then, how many humans? This is more tricky. Extrapolating from simple counts of individual human fossils is likely to have such huge error bars as to be useless. More subtle means would need to be improvised. One possible proxy might be the extent of obvious transformation of either strata (extent and thickness of the Urban Stratum, say) or of fossils (the proportion of the 'new' fossil pollen derived from our food crops to that of 'natural' pollen from pre-human forests and grasslands). The former method might be akin to that used by palaeontologists today, measuring to what extent marine strata have been disturbed by the action of burrowing creatures, though this reflects both the number of marine creatures and, as an independent variable, the speed of sedimentation, that is, to what extent the creatures have a chance to burrow through the sediment before more sand or mud arrives. The latter method is used by archaeologists to gauge the spread of early human agriculture.

Once a measure of some sort is obtained, there would then come the problem of extrapolating it over the rest of the terrestrial land area. And here appears a side-effect of the factor that in general will help to see evidence of humans preserved: on the whole, humans live more thickly in areas that have good preservation potential. An extreme example is the Nile delta (thickly populated; good preservation potential), contrasting markedly with the arid uplands farther inland (very thinly populated; low preservation potential). Extrapolating population densities from the preserved near-coastal evidence to the inland regions could thus exaggerate the numbers of people, and make the human takeover of this planet seem even more extreme than it is.

This exactly parallels the relatively good picture we have today of lowland dinosaurs, and our contrasting ignorance of what lived in the Jurassic uplands and mountains. To establish better constraints in both cases therefore needs more thought, and more evidence. Both humans and dinosaurs rely, directly or indirectly, upon primary plant productivity, which is climate-related. Thus, if one finds indications of inland aridity washed or blown into the biologically rich near-shore strata (such as impact-frosted windblown 'millet-seed' sand grains) then one might deduce that inland there was low biological productivity, and therefore lower populations of top predators, whether human or saurian. It is slow work arriving at such ever-clearer assessments of the deep past and its palaeobiogeography (but it does have its fascination).

Then, how about the distribution of humans: spread evenly across the land or clustered into agglomerations? Here one signal should be clear. There will be a marked distinction between the striking, if variable rock type of the Urban Strata (the remains of the megacities now seen so strikingly on Google Earth) and the strata representing the landscape in between, now mostly agricultural land. The latter will effectively be a fossil soil, and one that may be hard to distinguish from that produced on a natural landscape: differences in texture may well be preserved, resulting from ditches and other drainage or irrigation channels, and there will be the fossil content, particularly the pollen. There will be some intermediate types of strata, those representing low-density suburbs and scattered homesteads. Nevertheless the Urban Strata and the countryside-strata will be distinguishable: our future explorers will, indeed, be able to make geological maps of their distribution using their equivalents of hammer, base map, hand lens, and notebook. From that exercise an estimate of biological scale and density can be judged, just as geologists do today when mapping fossil coral reef deposits.

Our cities are now *big*, and growing ever faster. Los Angeles, Shanghai, Lagos, Amsterdam—their urban areas are now measured in thousands of square kilometres. The evidence that will be visible of them in the far future will be random slices, as their strata, after burial, are uplifted and tilted in arriving back at the Earth's erosional surface. These slices will be the surface outcrops of the Urban Strata, the lengths of which (as ascertained by geological mapping) will range from kilometres to tens of kilometres. To reconstruct their original area our chroniclers would need to make sensible extrapolation—to predict where any particular Urban Stratum would project to underground, assuming as a first guess that the cities were originally broadly circular in outline. They might also try to verify this, say, by drilling boreholes, for the distinctive human-made rocks should be easily recognizable in borehole core specimens.

By whatever means, the cities will be seen as human-built agglomerations each capable of containing, in some way, many *many* people. Because the evidence from modern times will overwhelm, in bulk and scale, evidence of humans in historical times, it may well be deduced that humans always were hyper-social animals, with a marked tendency to cluster. The pattern for most of human history, that is of most people living within thinly dispersed rural populations, will leave much less of a geological footprint. It is another way in which the geological record is skewed: the narratives it contains need to be unravelled with a watchful and sceptical eye, one open to alternative possibilities.

Nevertheless, how many people in a city? This is where the nature of the evidence (what is left of the cities) needs to be integrated with reasonable guesswork (how much living space in such an agglomeration would each individual human need?). The evidence will be heavily biased in favour of the underground. Underground pipes and pilings and tunnels are somewhat more likely to be preserved than cellars, and cellars are much more likely to be preserved (and preserved in much better shape) than ground floor rooms and buildings, while any constructions higher than this will almost always be in the shape of a mass of weathered and fragmented and often wave-washed rubble, and the whole will be compressed, cemented, and altered. The possible existence of skyscrapers will only emerge from detecting excessively thick rubble layers above the better-preserved lower city levels, and deducing that they must have formed from constructions that were very big and very close by. Far-future palaeontologists will have their work cut out for them even more than is the case for modern-day archaeologists.

Nevertheless, the basic modular structure of city structure should emerge, and it will be possible to work out, here and there, the dimensions of rooms and buildings—at least of their outline in plan. What were these enclosed modular spaces used for? Obviously for activities to do with life, given that organized death assemblages—cemeteries—are likely to be recognized elsewhere at some point in the excavations. The idea that the modules are some sort of highly elaborated shelter, a controlled and controllable living environment, should emerge by comparison with other terrestrial animals (that live in caves, or construct nests or burrows) or perhaps more by analogy with animals that cooperatively build highly elaborate constructions, such as termites and colonial wasps. The wide geographical range of humans, too, would be consistent with them finding means to live beyond their natural tolerances of temperature and humidity.

It would be natural to estimate, as a first approximation, that there would be one module per human, and this approximation might provide a good way to refine their estimations of the size of former human populations. At a slightly more basic level, this is how palaeontologists today tend to interpret fossil graptolites, mentally placing one zooid in each individual tube of the whole colony. Of course, this is not necessarily how it was with graptolites. There may have been several zooids in each tube, or different numbers of zooids occupying tubes at different stages of a colony's growth, or each zooid may have been able to move freely between component tubes; it may be that different graptolite species came to different arrangements. Nevertheless, it is standard practice to draw the simplest conclusions first, in the absence of other evidence (Occam's razor may well have a counterpart in societies very different from ours). In the case of the preserved human colonies, further work would show that modules (rooms) could be of greatly different sizes, as could modular structures (buildings). This would therefore require some rethinking of any original interpretation linking humans to the modules that they constructed. One step forward, two steps back.

Let us stay with the basics for now, though. Gigantic, complex city-structures, whether for termites or for humans, need to be engineered to arrange for food and energy to come in and for waste products to go out. It will become clear during the excavations that all the nutrients necessary to sustain such a concentration of city-dwellers could not easily be generated within the city, even if all of the available space outside the modular structures was devoted to some form of primary production. More likely, the food came in from the modified landscapes surrounding the cities. The deduction that humans were in some way selecting, modifying, and encouraging the growth of the plant matter that they required for food would also seem, if not inescapable, at least consistent with such an interpretation.

Would there be any direct evidence of these putative foodstuffs as fossils, beyond the ubiquitous pollen? Food being inherently perishable, this might seem one of the less likely pieces of evidence to be smuggled through to the future. But plant macrofossils are not uncommon generally in the fossil record—think of the fern fronds preserved in the Coal Measures, for instance. Stalks and ears of wheat, barley, maize, millet, and other food crops will almost certainly be swept into shallow pools and lakes and estuaries adjacent to farmland, and are no less capable of being fossilized than unmodified plants. Would they be recognized as cultivated? This might take some competent fossil plant taxonomy, though the fossils over wide areas may show little variety, reflecting the monocultures of modern farming, and will differ in form from their natural relatives. Cultured wheat, for instance, differs from its Middle Eastern ancestor, einkorn, in possessing larger grains that cling more strongly to the stem (thus allowing effective harvesting). They would be hard to recognize as human-cultured food crops on the basis of accidental finds—but then again, our excavators may be looking in the Human Event Stratum for just such evidence, and that would be a quite different thing.

What about processed foods? The preservation potential here varies, particularly where the food is canned or in jars or bottles. Here, it is the containers that may be fossilized to varying degrees, while their contents, if preserved at all, would be reduced to carbonized organic residues. Distinguishing food from other organic-based contents (shoe polish, say) would likely be difficult. Future palaeontologists, though, will be aware that a considerable part of their studied organisms' lives would have revolved around nutrition, so food (and not shoe polish) would be a sensible default option for such specific trace fossils.

There should be evidence of predation, too, to go with the herbivory. There have been many more animals consumed by humans than there have been humans, and so more potentially fossilizable skeletons of a relatively small variety of terrestrial animals—pigs, sheep, cows, goats, chickens—and a somewhat greater variety of fish. Although not all human-predated bones are discarded (a proportion being converted into, say, bone-meal) sufficient numbers of such bones should still turn up as fossils. Their taphonomy will be sharply distinct from that of the articulated and geometrically arranged skeletons in human cemeteries, and also from those of fossilized, predated skeletons of pre-human times. The predated bones will be disarticulated and butchered, very often mechanically sawn, scattered randomly or piled in midden deposits. This link in the food web should be quite clear, and our chroniclers may form an opinion on its suitability for an intelligent being, depending on their own sensibility and ecology.

After food, mode of reproduction generally figures among the characters that palaeoecologists try to interpret in an organism. The future Earth-explorers will soon become aware of the distinctive mode of reproduction of animals on this planet, with most animal species (and all mammals) having male and female forms, that are sometimes easily distinguishable on external appearance (such as lions) and sometimes cryptic (such as hedgehogs). It will probably be assumed that the human species possessed two sexes. But, in humans, the differences between the sexes are slight—at least to an impartial observer—and mainly expressed in soft part morphology. The skeletal differences, as regards hips for childbirth, say, may only be recognized once our chroniclers become reasonably experienced mammalian anatomists.

Reproductive mode is linked inextricably with the social structure and the rearing of the young. We can assume that animal populations of the far future will be generally similar to those of today, at least in the way that today's terrestrial vertebrates (or perhaps, more precisely, those that existed on the Earth up to ten thousand years ago) *generally* resemble the dinosaur-dominated assemblages of the Jurassic and Cretaceous. That is, they show a variety of types of social groupings (herd-forming organisms such as wildebeest and stegosaurs, essentially solitary ones like hedgehogs and—perhaps?—tyrannosaurs) and family structures (harems, monogamy, polygamy). The vast conglomerations of the Urban Strata would strongly suggest herding. Working out the nature of the family structure, though, will be trickier. Families will rarely be preserved *together*, so evidence might need to be sought in the preserved remains of the dwelling structures. A working initial assumption might be that each structure might house one family. As more of these structures are excavated, and reveal a great range of size and arrangements, from small, isolated constructions comprising just a few modules, to much larger ones with many, then this hypothesis might be dropped—or replaced by one suggesting great variation in family size and structure. Some questions are stubbornly resistant to enquiry.

Others, though, are more amenable. If the structure of families will remain enigmatic, their success in their main task—care of the young—can be more effectively assessed, by examining the age structure of the fossil assemblages. If the human remains are dominantly of mature specimens, then this shows that most of the young grew to maturity, while an abundance of immature skeletal remains suggests high mortality in infancy and youth. There are the usual sources of error and uncertainty. We can work out age structure via measurements of size or thickness of equivalent bones between individuals. Such measurements, though, would also include the effect of natural variation, in that some individuals are taller or more strongly built than others, even if they are exactly the same age.

Nevertheless, the analysis of skeletons from a fossilized cemetery from one of today's major cities should demonstrate marked success in steering young individuals through to maturity. An equivalent cemetery from the Middle Ages would show less success, in that most people died relatively young. If enough discoveries were made, the data would ultimately comprise fossilized human remains representing the last couple of centuries (a short time of very many individuals, often living long lives) and the several thousand years before that (a much longer period, with many fewer people, of whom only a few reached their three-score years and ten). It would also reflect geographic variation (at any time, young people in some parts of the world are more likely to survive than those in others). The picture, as it is assembled by the excavators, may therefore appear puzzlingly mixed. It will be drawn up also, of course, from the age-structure evidence yet to accumulate, evidence that will truthfully reflect the outcome of our own generation's hopes and fears for the future.

A picture could then be built up of intelligent organisms living extended lives, *en masse* in extremely high concentrations in vast manufactured structures. To build and maintain these, and provide continual inflow of food and water and raw materials, needs enormous, sustained inputs of energy. Human muscle power is a factor, of course, not to be discounted. It did, after all, we know (but can scarce believe), build the Pyramids. But once our chroniclers realize the sheer scale of the agglomerations they excavate they will be speculating as to where the extra energy came from. As intergalactic voyagers they will be acutely aware of the benefits of effectively harnessing power, and be equally aware that this single capability likely lay at the heart of the enormous, sudden success of this one species.

What were the energy sources? The most natural, nearest, and most constant source is the sun, the source of all the energy that flows through the Earth's web of life. This is not difficult to collect, except that large collectors are needed to harvest large amounts of sunlight. Then there is the energy locked within atoms, which will be quite familiar also to any space voyagers, and that is also capable of being harnessed, with sufficient care and skill. The future explorers might be looking for evidence of the use of such obvious sources, and perhaps being puzzled by its apparent scarcity.

An energy source that might be rare by galactic standards is deep-lying strata composed of black, fossilized sunlight, for that is in effect what coal, oil, and gas are: carbon harvested from the air by organisms that use the solar radiation to break open carbon dioxide and water molecules and then to construct hydrocarbons, that are then buried in layers—or segregated into underground pools—by the constant and highly specific action of the Earth's strata machine. It may be counter-intuitive and a little bizarre, from a true outsider's perspective, to consider the extraction and oxidation of compacted fossils as a means of powering a civilization.

Yet, there would be evidence of precisely this bizarre mode of energy collection. While the surface expression of the hydrocarbons and petrochemicals industry will be no more and no less preservable than any other construction—that is in fragmentary and altered fashion—it has deeper roots. Most of the Human Event Stratum (that includes the Urban Stratum as its most extreme form) will be a thin and patchy, albeit globe-encircling layer. But, the search for coal and oil has driven far underground, and has left tracks in a realm where preservation is guaranteed, at least until the tectonic escalator takes them to the surface. The indelible punctures in the crust made by humans in their search for hydrocarbons will be eye-catching trace fossils when encountered: roughly cylindrical zones up to a metre in diameter punching through the strata, often injected with barium-rich mud and stray hydrocarbons, and here and there metal-enriched from the borehole casings used. Will they be recognized for what they are, or would they be interpreted as natural, perhaps as hydrocarbon vents akin to miniature volcanic pipes? The collapsed and shattered zones of worked-out coal seams may be more obviously interpreted as ancient mine workings: especially when, traced across some cliff or river section, they suddenly give way to a pristine coal seam.

Enigmatic structures perhaps, but their significance might be grasped, when set against the evidence of a perturbed global carbon cycle, of sudden global warming and sea level rise precisely synchronous with—as far as it will be possible to judge—the Human Event Stratum. We now analyse similar phenomena that took place in the geological past: the Toarcian warming in the Jurassic, for instance, and that which took place at the beginning of the Eocene Epoch. For these, we seek candidates from among, say, sea-floor methane hydrates, for the vast quantities of carbon that must (as the isotope changes show) have poured out at those times into the atmosphere. Our future chroniclers will have a ready-provided source of carbon to explain the human-coincident warming. Would they be able to deduce the nature of this carbon transfer? If so, they may wonder how seemingly intelligent, organized beings selected such a transient and dangerous source of energy to build an empire on.

Dangerous, yes . . . But the liquid hydrocarbon fuel in particular is *so* convenient and portable. And it provides (if they but realize) the means to answer another question to be posed by the excavators: the locomotory and migratory patterns of the human animal. Local migration would seem unavoidable, simply to supply nutrition to the centres of the city-agglomerates, just as is done by colonial wasps and ants. Given the scale of the human colonies, though, muscle power alone would not suffice. There must have been some form of assisted locomotion. How?

Unlike the scale of our cities, the sheer extent and length of the road network will be extraordinarily hard to reconstruct. Not because roads cannot be fossilized. Submerge and bury a well-constructed road under sediment: with its base of hardcore and its tarmac surface and its bounding kerbstones, it will preserve as well as any concrete cellar. Perhaps better, for the effects of sedimentary compaction will not be so destructive. But when exhumed it will be exposed *in section*, as a cross-cut revealed in a cliff or a crag. Part of the road will be a ghost road in the sky, eroded and removed forever. The other part will descend deep into the rock stratal mass under ground; it might be followed for a metre or two, with considerable difficulty, by excavating into the rock, but after that it will be effectively inaccessible. On the finding of such a structure, the first question will be whether the section represents a natural cut across a linear structure, or one across a small circular or square construction—some kind of small ground-level platform, perhaps.

It is a classic field geology problem: given a two-dimensional section through an object—what is its three-dimensional shape? Sedimentary rock samples often show circular patches of a different colour or mineralogy: are these spherical in shape, as in some kind of chemical concretion; or are they in fact sections through tubes, such as fossilized worm burrows? If the rock is soft, one can excavate into it to find out. If the rock is hard, then one looks for more sections; if the structures are tubular, then some sections across it will run more or less along the tube axis, and so be elongated in shape; others, cut diagonally across, will be oval in shape.

Roads being bigger than worm burrows, the geometrical problem is inherently more difficult to solve. Perhaps a plan view of a road fossil may be had, occasionally, where areas of entire bedding plane are exposed, as occasionally happens with dinosaur trackways. But trackway surfaces that approach, say, a football pitch in size are exceedingly rare. That such surfaces could run for miles, still less form networks extending across hundreds or thousands of miles, may be a wild hypothesis of one of the more fanciful and conceptually reckless of the future palaeontologists. But, it would be virtually impossible to prove, especially given that the roads will soon run into the erosional realm, and thus be removed without a chance of being buried, or segments will be eroded and destroyed after abandonment, but before the land surface across which they run is buried.

Today, for instance, the Sacramento River delta in California is subsiding rapidly, because of drainage of the rich farmland it provides. When it floods, roads are submerged—but sections of them are also washed out by the floodwaters, especially around where the levees are breached. As sea levels rise, flood frequencies will increase. Eventually, the roads will be submerged for good, but the network that will be buried under sediment will be fragmented and partially eroded, and hence even more of a future geological puzzle to reconstruct.

The wheeled transport machines that now run in such numbers along them may also fare rather poorly as regards long-term preservation. Were they made of ceramic, concrete or bone they would fossilize, perhaps even rather well. But iron and mild steel easily rust at the surface, and corrode and dissolve in the chemically reducing conditions of burial, while the compaction would crush the structure as effectively as the jaws of a breaker's yard; rubber and plastic would carbonize, and glass devitrify. It would take some more than averagely good preservation to discern that there were rotating wheels, and yet more to show that their rotation carried the whole contraption along the ground surface.

Nevertheless, there would be the strongest of circumstantial evidence for migration, and for long-range, intercontinental migration, at that. First, there would be the sheer geographical extent of humans, on every continent, including those that would originally have been widely separated by deep oceans (this is assuming, of course, that our aliens would by then have divined broadly how the continents were arranged in human times). Humans must have crossed those oceans, somehow.

There will, too, be fossil evidence of that unique mass cross-migration of other species, both animal and plant, both terrestrial and marine, at exactly the time of the human phenomenon. Emerging palaeogeographic reconstructions will rule out one possible explanation: the existence of a single supercontinent in the Human Period. The emerging rediscovery of plate tectonics will also preclude the former existence of a network of trans-oceanic land bridges. One conclusion will seem, if not inescapable, at least the most likely of the hypotheses to explain this phenomenon: that these animals and plants were somehow caught up in the travels, as humans not only spread across continents, but travelled backwards and forwards between them.

There are a limited number of means to achieve such travel: through the water, over the water, or in the air. What chance fossilized shipwrecks (of submarines or boats), or the remains of crashed aeroplanes? On the one hand good, in that these would fall generally into the preservational realm of marine sedimentation. However, these occurrences will be isolated, while the fate of the *Titanic*, now corroding fast, suggests that many such remains may be in quite poor shape by the time they become incorporated within strata. Some of the rubbish tossed overboard—bottles, discarded cups and plates—may provide a clearer indication of humanity's contact with this realm.

So, a mobile society, and thus technologically advanced. But to what extent? The trappings of the modern age, in many ways, seem less dur-able than the more primitive tools of our forebears, and many are so large (an ocean liner, say) that excavation through the corroded, flattened remains of such a find would be an enormous undertaking. But the miniatures of human manufacture are now produced in such abundance that some might just find their way across the abyss of time. Laptop computers, mobile phones, MP3 players, digital watches are now bought in their many millions and discarded like leaves from an autumn forest. Each, though, is a miniaturized gem of precision manufacturing, almost as finely wrought and complex as a living cell, yet obviously fabricated. And as perishable as tissue, almost, once discarded to the elements. But they are common now, these playthings. If discarded somewhere amid the debris of a city where, say, lime is rapidly crystallizing and hardening, or a pool of thick oil is turning to sticky viscous tar, or pyrite is beginning to crystallize in stagnant harbour mud, parts of their tiny filigree mechanisms might just become entombed and protected, like flies in amber.

In 1900, sponge fishermen came across a 2 000-year-old shipwreck in the Aegean Sea, off the islet of Antikythera. They recovered rich jewellery, statues, coins, glassware. Among the spectacular haul was a lump of bronze and wood, about ten centimetres across. This lay unnoticed in a museum for a few years; then, once thoroughly dried, it happened to crack open, to reveal the remains of toothed gear-wheels. This aroused some interest, but the object was fragmentary, and had been severely corroded by the seawater. It was only from the 1970s that X-rays, and then computer-aided tomography, revealed the structure of the Antikythera Mechanism. This showed that the Greeks had built an astronomical computer of some 32 interlocking gear-wheels. It calculated phases of the moon and the solar calendar, predicted solar and lunar eclipses, and more. It is a staggeringly intricate and cleverly designed machine. This was technology invented and then lost, for nothing of comparable sophistication was to emerge (as far as is known) virtually until Renaissance times.

Maybe, amid the fossilized ruins of the Human Empire, ruins that imply large-scale earth-moving and organized construction, such future Antikythera-like, paradigm-shifting mechanisms may be found in their own particular *Lagerstätten*, finely preserved vestiges of our electronic wizardry. From an alien perspective, they may not be interpretable as regards use and purpose, in all likelihood, but nevertheless they would suggest a capacity for sophisticated construction and delicate manipulation.

There is one place, though, where pristine, complex human artefacts will lie for a near-eternity, protected from decay and corrosion, and yet available for easy inspection to those that find them. The artefacts that currently lie there, true, are a little *démodé* by contemporary standards, but they will provide powerful evidence of human technical capabilities—and of at least one mode of human transport. On the Sea of Tranquillity, some 250 000 miles away, the thin lunar soil supports the small mass of a reflector for laser range-finding, and some seismographs built to detect moonquakes. On that still, absolutely dry landscape, these, and the scattered footprints around them, and the artificially stiffened US flag, and that golf ball fallen a little in the distance, should be quite unaltered, a hundred million years from now, except for a little surface pitting from micrometeorites. There are six such lunar landing sites from the Apollo missions. On the vast surface of the Moon, they are a few needles (although exquisitely preserved ones) in a haystack of barren rubble. Given the treasures to catalogue on Earth, though, the future explorers may devote scant time to a dead and prosaic satellite.

On Earth, the discovery of complex manufactured objects—half a wristwatch, perhaps, or part of a CD player's mechanism—would immediately suggest a capacity for division of labour (someone has to provide the food for those who manufacture the technological marvels). It would mean that many of the modular compartments of the city-constructions were manufacturing-places, and not just shelter-places (and *that* would mean modifying those population models). And, for our chroniclers, it would be evidence of the importance of communication for the members of the Human Empire: to include also communication *through* time, to transmit detailed instructions and invented blueprints across the generations.

Animals today signal to each other vocally, or through visual signals, or through touch, or even through smell, as in the kaleidoscopic olfactory spectrum of the bloodhound's brain. Some are blind, some deaf, some possess a sense of smell even worse than do humans. It would be hard for our excavators to find clues as to which of these would have been important to humans in communication, from our preserved bones or from our artefacts. A visual sense, perhaps from fossilized statues (though these could have been interpreted by blind individuals through touch, just as does one of today's most celebrated zoologists, Gerhaard Vermeeij, who has been blind since infancy: he studies molluscs, and works out their taxonomy and ecology through touch alone). The peculiar code we have of writing? Books, alas, do not resemble clay tablets in terms of preservation potential. A *Lagerstätte* of books with preserved pages, somehow left flexible and uncarbonized by burial, is hard to envisage. Even if fragmentary letters were found, perhaps carved into stone (were they even to have been inscribed by the legendary Kilroy), or on coins, it would be hard to tell such messages from, say, decorations, things of a pleasing shape but possessing no specific meaning.

At this level of interpretation, when it comes to establishing what humans were like—as people, in effect—the clues may run into the sand. There is unlikely to be a Rosetta stone to allow, from an alien perspective, such understanding. The bare bones of organization, of a level of technological sophistication, of mobility, yes. But it is hard to think how the normal workings of geology and taphonomy can capture anything that one might describe as embodying the essence of humanity. Not knowing the extent to which humans could even hear sound, what chances are there of divining the existence of such a thing as music: sounds made harmonious for pleasure, and to inspire emotions more complex than pleasure? There is still less chance of guessing at the creations of a Mozart or a Schubert (or an Ellington or an Armstrong): immortals, yes, but immortal only for as long as human ears are there to interpret them. Their music cannot long be fossilized; even if petrified fragments of LP or CD lie somewhere among the city rubble-stone, their freight of melody is unlikely to be revealed, even by the closest of scientific analysis. The preserved and time-transmitted fragments of a written language are likely to be pitifully fragmentary, too, so what chance then of our explorers even considering the possibility of a Goethe, or of a Shakespeare? Or, if analysing carved or engraved representations of the human form, of divining the motives that drove Michelangelo or Rodin—or caused a child playfully to mould some clay? Or, in observing the courtship and mating behaviour of terrestrial animals, predicting the many finely shaded expressions of human love (and hate) from Dante to Byron to Gershwin, or of the multi-layered ties that bind friend and family and colleague.

Analysing the needs and drives of an alien civilization that is preserved only through relics, could one interpret, say, independently, the scale and complexity of the human need (once fed, once sheltered) to be stimulated, pleasured, entertained, inspired? For this need now drives perhaps the greatest single industry on the planet, of spectator sport and mass tourism, Hollywood and Bollywood and television, and paperback fiction and poetry and computer games and opera houses. Would that aspect of humanity emerge in the analysis? Frivolichnia are there in plenty—but will probably be misinterpreted wildly; it would likely take the Zaphod Beeblebrox variety of intergalactic explorer to recognize them for what they are. A stadium is for sport, not for communal discussion and decision-making. A bullring is frivolichnia (for some), not fodinichnia. There's nowt so strange as folk, but strangeness doesn't fossilize, by and large.

Somehow connected to one end of this scale, humans also seek meaning out of short lives. Many hope and believe in some divine purpose behind the Universe, and religions have arisen in virtually all human societies. But preserved fragments of mosque and church in the Urban Stratum would surely be classified together with remains of school and leisure centre and factory floor, as larger complex constructions of uncertain purpose.

Nevertheless, the data, as it accrues, should give clues to both constructive and destructive drives among humans. Some part, at least, of humanity's inner contradictions should be impressed, as it were, into the strata. The evidence of widespread physical care for many long-lived individuals (with healed bone fractures, say, or carefully repaired teeth) will contrast sharply with the signs of intra-species violence when the first mass grave of skeletons is discovered, and interpretations of natural catastrophe would not square with evidence of unhealed puncture and cut marks in breastbone and skull.

Would the scale of the trace fossils of murder, of these killichnia, be realized, and their industrial-scale production be surmised? The realization may or may not come rather late that a good deal of the resources of this civilization were devoted to creating machines that puncture, or slice, or fry other humans. Again, depending on the sensibilities of our excavators, such production might be interpreted as representing anything from standard competitive practice through to horrific barbarity.

But this is a fairly simple opposition of care versus murder. The Occam's razor of scientific study will undoubtedly miss the richness and sheer perversity of humanity. In dealing with memories and relics and interpretations, Keith Douglas, that poet informed (and then killed) by the arbitrary narrative of desert war, asked that he be remembered when he was dead, and simplified when he was dead. It is one thing to unearth some indication of the scale of a terrestrial phenomenon, and reconstruct its shapes and patterns (even to a fine level of detail), and to glean something of its context and relations within the Earth system. It is quite another to understand it.

One might refer back to those graptolites again. After two centuries of study, thousands of species are known (six hundred and ninety-three in Britain alone, at the last count). They were successful members of the plankton, a familiar enough type of community to biologists, for a hundred million years or so. They have beautiful and intricate and quite species-specific skeletal shapes, that are sometimes so well-preserved (if acid-dissolved out of limestone or chert, for instance) that they might be modern biological specimens, preserved down to molecular level and interrogatable by electron microscope. There are a quite limited number of functions and needs that must have influenced the particular shapes that were evolved: feeding, reproduction, hydrodynamics. And yet, to a quite frustrating degree, we fail to understand how function and need and form fitted together. Commonly, for instance, these creatures evolved thorn-like spines in particular arrangements: did these have a hydrodynamic function, to slow the rate of sinking?—or were they there to help feeding (by allowing the zooids to crawl out along them to fish in more distant waters)? And why did they build (and devote considerable resources to) those tough, substantial skeletons at all? Nothing else in the zooplankton has anything quite as resistant. They could surely, on the face of it, have managed with an armour that was altogether more lightweight. Yet they didn't—why?

Now translate such questions into the many different forms and functions and needs (and desires) that must arise in an organism intelligent and sophisticated enough to manipulate the environment around it and to build a gigantic and intricate material culture. The criss-crossing, multi-dimensional array of interpretive possibilities now balloons out into, for all practical purposes, infinity. Then try to pick among those possibilities from an extra-galactic perspective (remembering, from Stonehenge, how difficult the search for understanding may be even from the perspective of members of one's own species). One can attempt to address the generalities of the problem, as we have done above, and analyse humans and their constructions in broad ecological terms. But what lies at the heart of humanity is probably not available for preservation, even in the finest

Lagerstätten. It may, though, be of some comfort to know that we, the people, are constructing an eternal cosmic puzzle, the most mysterious by far for many trillions of miles in any direction.

There is a flip side to this long-range view of humanity's legacy, though it is one that contains less comfort. The sharper and clearer is the message that we leave for far-future explorers to discover, the more damaging are the effects for ourselves. The clear stratigraphic message of multi-species extinction, global temperature spike, and abrupt sea level rise will guide these explorers to the petrified cities. It will also be a sign of unbearable stress upon a species that single-handedly took control (using that word somewhat loosely) of the levers of planetary regulation. The deeper the footprint that we leave, the greater will be the immediate calamity that awaits our children. This particular fossilization process has harmful—indeed, potentially catastrophic—side-effects.

It would be immortality cruelly won. Best to leave as small a message as we can, to impress today's human footprint as gently as possible into the strata of tomorrow: to diminish, in as far as we can, the stratigraphic signal that we leave behind us. At the time of writing, fine words encourage such restraint, but the practice is less convincing. Carbon emissions, extinctions, habitat loss, sea level, and human population are all rising, and if they continue to rise for more than a generation, then prospects will be bleak indeed. Yet it may not be too late, still, to change course from a path that, in producing a signal that will remain clear for many millions of years, promises misery or untimely death for billions of people.

What to do? A clear prospect of direct action, perhaps, and directed application of substantial resources (a modest proportion of the world's military budget would go a long way) to smooth out the bumpy ride that looks to be in store for the human race. The adoption of softer energy paths. A re-foresting of the world. Finding contentment without the compulsive overuse of resources. Perhaps most importantly, curbing population growth—though without the natural selection of the strongest and most ruthless. To achieve sustainable numbers while still curing the ill and succouring the weak would be our civilization's greatest triumph.

Perhaps technology will rescue us. Another century of accelerating human capabilities in computing, in nanotechnology, in biotechnology might transform our ability to mitigate, or adapt to, the changes reverberating through the Earth. This alone, though, might be insufficient. The current inequalities in wealth (and thus in health, and lifespan) between the hyper-rich and the dirt-poor may seem to any visiting alien explorer to be the stuff of science fiction. They will certainly make it difficult for us to act collectively as our drama unfolds. So steps towards encouraging a common humanity might help preserve humankind, rather than merely preserving its remains.

That can seem a distant prospect at times. Yet conserving living organisms is far more important than conserving fossils (and here one speaks as a lifelong palaeontologist). The Earth, in sustaining and harbouring these organisms, is by far the most intricate, the most subtle, the most complex and valuable object in space for many, many billions of miles in any direction. It would be not merely an Earthly disaster if its surface was converted to the kind of wasteland that appeared after the Permian–Triassic or Cretaceous–Tertiary boundary extinction events. It would be a cosmic tragedy, one in which the injuries sustained would not heal for millions of years. This is a denouement that we should strive to prevent, while we are still able to.

But, whatever we as a species do from now, we have already left a record that is now indelible, even while the scale of this fossilization event is still in question, and within our power to determine. Humankind has, through its various activities, done enough to preserve its relics into the far future. The 'environmental' changes that we have set in train will, without a shadow of a doubt, be translated into the solid rock of the Earth. The Urban Stratum is now, in substantial part, effectively eternal. More: our actions now will literally be raising mountain belts higher, or lowering them, or setting off volcanoes (or stifling them), or triggering new biological diversity (or suppressing it) for many million years to come. The knock-on effects of our geochemical experiments are unpredictable in detail, but will be substantial and likely surprising. We have left our mark. However we are interpreted in some distant future, there will be little doubt that we will be associated with—and responsible for—some of the most extraordinary geology of this, or any other, planet.

Further Reading

van Andel, Tjeerd. 1994 (2nd edn). *New Views on an Old Planet*. Cambridge University Press. Lovely account of the Earth that elegantly interlaces tectonics and climate.

Beerling, David. 2007. *The Emerald Planet*. Oxford University Press. Very nice account of the Earthly role of the (all too often) forgotten kingdom of the plants.

Broecker, Wallace. 1985. *How to Build a Habitable Planet*. Eldigio Press. It's a text-book, and hence more demanding than most on this list, but it's worth it. Written by one of the greats of ocean research, it goes from the workings of stars to the controls on climate.

Cadbury, Deborah. 2001. *The Dinosaur Hunters*. HarperCollins. The topic has had many books written about it, but this account came as a (highly pleasurable) surprise to me in the freshness of its approach.

Darwin, Charles. 1845. *The Voyage of the Beagle*. Wordsworth Classics. A remarkable journey around the world, narrated by the greatest of all natural historians.

Dawkins, Richard. 1986. *The Blind Watchmaker*. W. W. Norton & Co. My favourite among this prolific author's takes on evolution.

Diamond, Jared. 1997. *Guns, Germs and Steel*. Vintage books. Fascinating if bleak examination of just how humans took over the Earth.

Fortey, Richard. 2005. *Earth: An Intimate History*. Knopf/Random House. Excellent on the various phenomena—volcanoes, earthquakes—that are by-products of the working of Earth's plate tectonics engine.

Richard Fortey's *Trilobite!* (2000, HarperCollins) and *Hidden Landscape* (1993, Pimlico) also get nicely under the skin of strange ancient fossils and landscapes, respectively.

Gould, Stephen Jay. 1978. *Ever Since Darwin*. Burnett Books. The first of Gould's many fine essay collections that I ever read, and still (for that reason?) my favourite. Of his longer books, *Wonderful Life* (1990, Hutchinson) is remarkable in basing a splendid—and best-selling—narrative on the dry and technical content of three palaeontological monographs (and a philosophical alternative view of the same fauna is *The Crucible of Creation* by Simon Conway Morris (1998, Oxford University Press)).

Imbrie, John and Imbrie, Katherine. 1986. *Ice Ages: Solving the Mystery*. Harvard University Press. A great account of the early scientists who discovered the reality of past glaciations, and of their successors who worked out the astronomical mechanism behind them. It's written by one of the pioneers among the latter and his journalist daughter; they make a fine team.

Kunzig, Robert. 2000. *Mapping the Deep: The Extraordinary Story of Ocean Science*. Sort Of Books. Marvellous. As good on the structure of the oceans as on their inhabitants.

Leakey, Richard and Lewin, Roger. 1996. *The Sixth Extinction: Biodiversity and its Survival*. Weidenfeld and Nicholson. Fine, thought-provoking account of extinctions, past and present.

Lovelock, James. 2006. *The Revenge of Gaia*. Allen Lane. By the creator of a distinctive and influential interpretation of the Earth, and an eloquent description of its (and our) possible future.

Moore, Ruth. 1962. *Man, Time and Fossils*. Lowe and Brydone, London. A book that tells the stories of the people who put together the story of biological evolution, and of human origins. It enthralled me as a child, and my sister Anna, who gave it me for Christmas, little knew what she was starting. It is still one of the best-written accounts of this topic that I know.

Nield, Ted. 2007. *Supercontinent: Ten Billion Years in the Life of Our Planet.* Granta Books. Highly readable and nicely quirky account of the working of past (and future) plate tectonics.

Pratchett, Terry, Stewart, Ian, and Cohen, Jack. 2000. *The Science of Discworld*. Ebury Press. There are fables (Pratchett's for one, Miyazaki's films for another) in which a subtle and intelligent view of the world underlies the magic and adventure. This is a hybrid of the Unseen University and of alternative worlds, and works very well.

Redfern, Martin. 2003. *The Earth: A Very Short Introduction*. Oxford University Press. Excellent on the inner workings of the Earth (and far more comprehensive than one might expect for a book that fits nicely into a jacket pocket).

Roberts, Neil. 1989. *The Holocene: An Environmental History*. Blackwell. A textbook on the environment of the last eleven thousand years—yet highly readable.

Rudwick, Martin. 1976 (2nd edn). *The Meaning of Fossils*. Neal Watson Academic Publications. A scholarly yet highly accessible account of the beginnings of palaeontology. It shows how hard it was really to divine, in those early days, that a fossil was once a living thing. A magisterial update by the same author is *Bursting the Limits of Time* (2005, University of Chicago Press).

Walker, Gabrielle and King, Sir David. 2008. *The Hot Topic: How to Tackle Global Warming and Still Keep the Lights On*. Bloomsbury. Our environmental legacy is something that is best kept to a minimum: here is a clear and persuasive account of how the heat might be kept down.

Wilson, E. O. 1993. *The Diversity of Life*. Harvard University Press. Life and what we are doing to it by one of the great twentieth-century biologists.

Winchester, Simon. 2001. *The Map that changed the World: William Smith and the Birth of Modern Geology*. Harper Collins. The story of the man who showed us how to read the Earth.

Index